COMBAT LOADS FOR THE SNIPER RIFLE

Ralph Avery

Desert Publications
El Dorado, AR 71731-1751 U. S. A.

Combat Loads for the Sniper Rifle

Published by Desert Publications
215 S. Washington Ave.
El Dorado, AR 71730
info@deltapress.com

12 11 10 9 8 7 6
ISBN: 978-0-87947-544-4
Printed in U. S. A.

Desert Publication is a division of
The DELTA GROUP, Ltd.
Direct all inquiries & orders to the above address.

Neither the author nor the publisher assumes any responsibility for the use or misuse of the information contained in this book. This material was compiled for educational and entertainment purposes and one should not construe that any other purpose is suggested.

Warning!!

The Publisher (Desert Publications) produces this book for informational and entertainment purposes and under no circumstances advises, encourages or approves of use of this material in any manner.

Table Of Contents

Dedicated to my wife Irene, whose patience, encouragement and typing made this book possible.

Introduction

Historically speaking, the art of sniping is very old. One can learn much about the art of ancient warfare by procuring books from Europe, notably from the British Historical Museum, on the weapons and cleverly designed siege instruments of Alexander the Great. This military leader employed specially skilled archers placed far above the ground to permit them to hit the leaders of a defending city behind its fortified walls. Some of the siege instruments were very ingeniously designed and were more than enough to make a person realize that much of Alexander's so-called "military genius" was in his ability to understand and maximize the potential of military engineering.

The longbow, made famous by the English, required a very lengthy period of training to be really skilled with it. However, a powerful man with a heavy pull bow, on which he had trained 10 to 15 years, was a very dangerous opponent indeed.

Moving closer to our own times, the crossbow occupies a special niche in the history of sniping. The crossbow was able to release in a fraction of a second, the stored energy that was placed in it by perhaps 10 to 15 seconds of intense effort on the part of the person operating it. Some crossbows have a pull of several hundred pounds and employ a winch for cocking. As a matter of historical interest, set triggers were developed to enable precision work to be done with the crossbow. The 500 to 800 lb. force delivered by the crossbow demanded some arrangement to permit releasing the bolt without disturbing the crossbowman's aim.

Penetration and range of the more powerful crossbow was phenomenal for the times. They were quite capable of shooting through a 2" solid oak door or penetrating an armored knight, plus giving a very serious wound to his horse. Even more important was the fact that effective use of the crossbow did not require several years of intense training. The rate of fire of the crossbow did not compare with what a skilled man with a longbow could do, however.

The introduction of the crossbow put the fear of God in the heavily armored cavalry of the day. As a matter of historical interest, one of the Popes in the Vatican, during the Middle Ages, issued an edict banning the use of the crossbow on Christians. The crossbow was a most useful tool for the individual peasant, also. In some ways, the situation then was frequently one of almost total breakdown in law and order. Renegade military deserters were always a potential problem and the crossbow gave the peasant a weapon to deal with such problems as they arose. As a matter of fact, I have long suspected that the reason European farmers lived in collective groups in small villages, rather than living alone on their land, was to have better protection against marauding groups of pillagers. It must be remembered that the means of communication of the day was limited to the speed of the fastest horse. Thus the individual farmer, along with the other villagers, could not get a warning days in advance of a roving group of military deserters or other hoodlums. Hence, their lifestyle had to be arranged to deal with emergencies in the most effective way on very short notice.

As we move along in history, the crossbow gives way to the muzzleloading black powder rifle. The so-called "Kentucky" rifle was perhaps the first example of the muzzleloading sniper rifle on this side of the Atlantic. One of the first examples of the military use of this weapon was in the Braddock Campaign against Fort Duquesne prior to the Revolutionary War.

The military exploits of George Rogers Clark utilized skilled riflemen during certain battles of the Revolutionary War. The Battle of King's Mountain was primarily decided by use of sniper rifles.

The Battle of New Orleans, during the War of 1812, saw the use of sniper rifles by troops serving under Andrew

Jackson. After the battle, the British were amazed at the range and accuracy of Jackson's sharpshooters. One tall Highlander's body was found 240 yards from the American breastworks. He had been hit in the head simultaneously by two American shooters who were well over 100 yards apart on the American lines.

The Battle of Adobe Walls some 60 or so years later featured less than 30 buffalo hunters against some 500 to 700 Indians. The buffalo rifles were quite capable of penetrating a horse's body at quite some distance and inflicting serious damage to the body of the Indian suspended on the horse's off side. Equally important, these Remington and Sharps rifles had a very high level of accuracy and their metallic cartridges gave them a fairly high rate of fire for the times.

Almost every student of the rifle is well aware that both the Union and Confederate Armies used snipers during the Civil War.

Then, as now, reliability in terms of hitting a target at long range was an absolute must. This requirement means accuracy and a flat trajectory, plus considerable wind-bucking ability are needed in the sniper rifle cartridge. Every bit as important is the necessity for reliable penetration by the sniper rifle projectile. Much of the testing in this book was aimed at evaluating commonly available types of rifle projectiles in terms of reliable hitting and certain penetration. The influence of velocity and bullet construction, as well as bullet mass, was investigated.

It is my hope that this book will be of value to the independent police agencies of this country and to those self-reliant citizens who are preparing to be able to defend themselves, their homes and their families in case of a localized or national catastrophe.

Cartridge Case Selection

The start of making the finest possible ammunition is the selection of cartridge cases that are the best that can be obtained for the purpose at hand. Combat loads for the sniping rifle have many things in common with any other reloads. The sniping rifle is supposed to be able to make a first round hit, and if necessary repeated hits, on a small target at a long estimated distance. It is supposed to do this under almost any conditions of wind and weather, provided a capable rifleman is firing it.

Cartridge case selection will greatly aid in getting the highest level of accuracy the rifle is capable of achieving. Selection of cartridge cases will have a very significant effect on the maximum safe velocity the particular rifle used can achieve. Lastly, intelligent case selection will aid in achieving the best possible reliability of the rifle when used in less than ideal conditions.

The techniques we are discussing have a great deal in common with those developed by 1000 yard competitive riflemen/reloaders. When you consider that a sniping rifle should be able to give a very good account of itself on a target range at 500 to 1000 yards, if it is going to amount to much as a sniping rifle, the loading procedures commonly used by 1000 yard riflemen make sense. The reliability of a sniping rifle should be on a par with a rifle you would carry in Alaska after brown bear at very short range. Therefore, diligent attention to cartridge case selection is clearly in order.

Cartridge brass is a material that can be softened by heat but can only be hardened by cold working it. The hardness of the brass is very closely correlated with its tensile strength and yield point. The yield point can be thought of as the stress level where the brass starts to flow or move. This is the action that creates loose primer pockets and similar problems. I have been a serious student of this variation in cartridge cases for well over 20 years. In 1968, I had an article published in *Gunsport* magazine on "Cartridge Brass", and in 1973, well-known writer Bob Hagel and I had an article on the subject in *Handloader* magazine.

In 1958, at the New York State 10 caliber championship matches, an incident occurred that started me off on a study of brass as used in cartridge cases. The result of this work is partially summarized in this chapter. The incident was a blown primer in the .30-06 handloads used by the man I was scoring for in the 200 yard Offhand match. The load was a normal one, using the 180 gr. Sierra Match King and Hodgdon 4895 powder. He had fired several hundred rounds in 1957 with no problems, using Den 42 cases. On this day, he was using the same load in '57 match cases, which are softer than the Den 42's. No more loose primers were encountered in that string.

Handloading articles that give some data on telltale indications of high pressure often mention that normal loads may give signs of high pressure in cases a good bit softer than normal. What constitutes "hard", "normal" and "soft" cases? Where do you draw the line? These are not easy questions to answer, but when sniper rifle users attempt to boost velocities in their pet cartridge or cartridges, some non-destructive test is highly desirable to give a reasonably accurate assessment of brass quality before loading. The critical point in a cartridge case is the head, or base. A weak neck may enlarge groups, a weak case head may put a few new holes in your head! This is not an idle theory. I have met two men who had factory .220 Swift loads blow up. One is partially deaf in one ear and the other had 80-plus brass particles removed from his eye.

We don't want problems like this in our sniper rifle combat loads.

Inferior brass definitely exists, just as "lemon" products exist in cars, dishwashers and other high volume-produced merchandise. I've spent too many years as a quality control and Manufacturing Engineer to have any illusions that inferior brass will not be made in the future.

In my opinion, chamber dimensions, brass hardness and permissible pressure are directly related. A minimum chamber, maximum case head size means little case expansion and very hard cases can be used with quite high permissible pressure. Conversely, a sloppy chamber that forces the case to expand should have a softer case to prevent rupture from localized bending. Factory engineers tell me that a rimless case can be constructed stronger than belted cases. This may well be true, since I have only once found belted cases comparably hard as Arsenal 7.62 NATO cases, Federal .30-06 and .270 Winchester cases and DWM .30-06 cases. Brass, as mentioned earlier, is a material that can be softened by heating, but not hardened. It can be hardened by cold-work.

My technique of testing cartridge cases is to hardness-check the solid head portion of the case just ahead of the extraction cannelure with a Rockwell "B" Hardness Tester in three places. This location is easily checked and is also the portion that keeps gas from your face.

Rockwell Hardness Testers are routinely used in machine shops and heat-treating plants. The "C" side is for tool steel and similar hard materials. The "B" is the one of interest to the handloader.

The graph of hardness, elongation, tensile strength and yield strength shows the relationship of these values. Where do we draw the line on hardness values? A yield strength of 50,000 PSI means about 76 Rb hardness. Since peak pressure of a load is somewhat greater than copper crusser value, this hardness of case would seem adequate for 45,000 to 47,000 PSI, or in other words, mild .30-06 loads. Brass Rb of 76 and below should, in my opinion, be considered soft and should be rejected for use in the sniper rifle. Rb of 77 to 81 should be considered normal, Rb 82 to 87 would rank as hard and

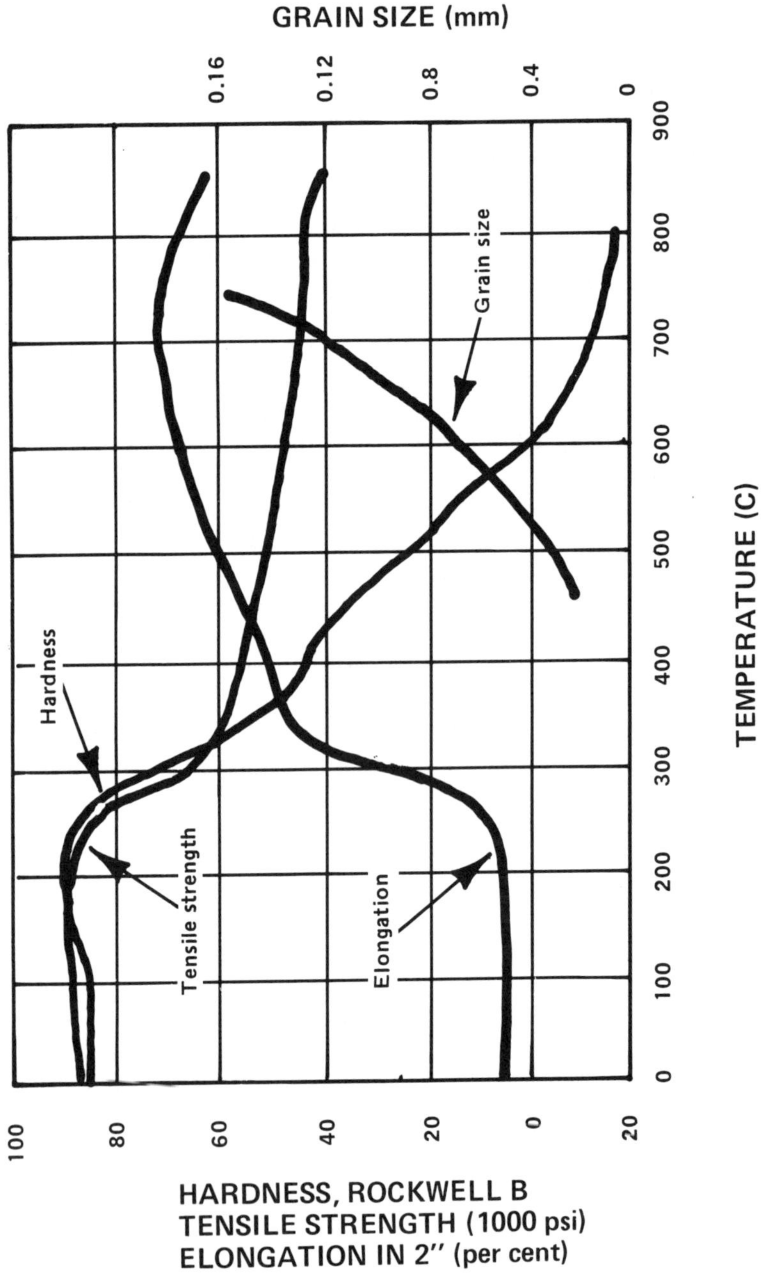
GRAIN SIZE (mm)
0.16
0.12
0.8
0.4
0
Grain size
Hardness
Tensile strength
Elongation
0
100
200
300
400
500
600
700
800
900
TEMPERATURE (C)
100
80
60
40
20
0
20
HARDNESS, ROCKWELL B
TENSILE STRENGTH (1000 psi)
ELONGATION IN 2" (per cent)

ELONGATION, YIELD & TENSILE STRENGTH
1000 psi – ROCKWELL

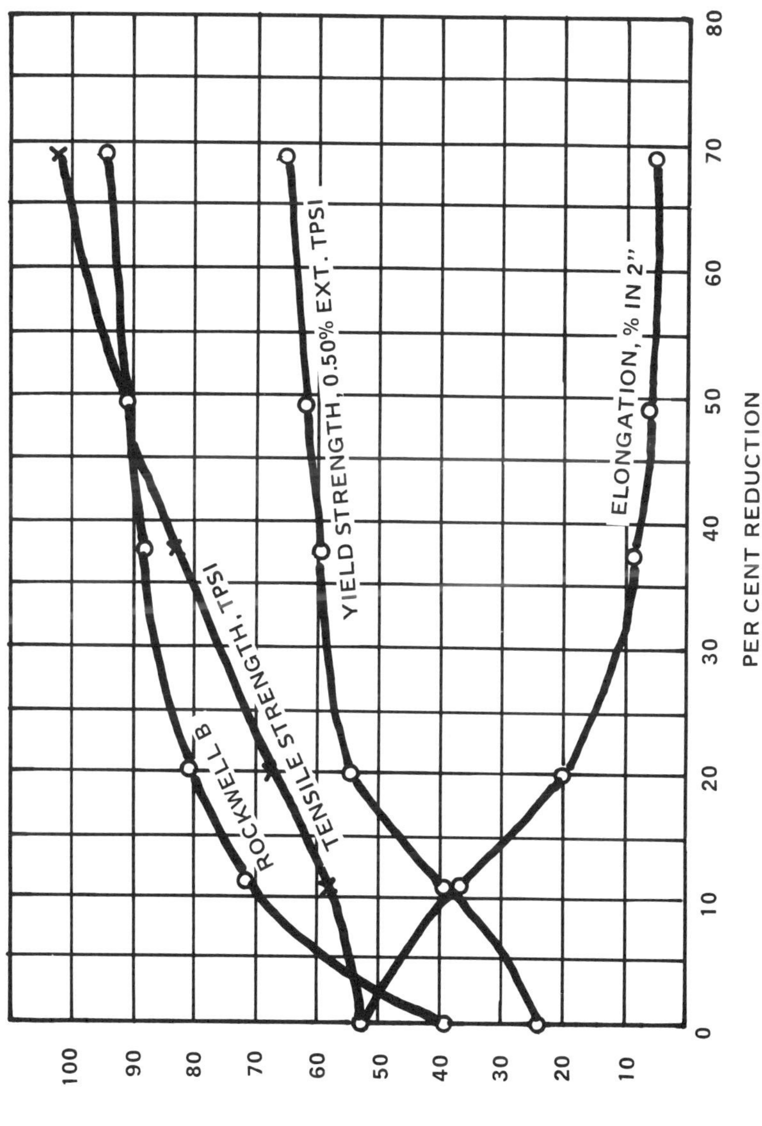

CARTRIDGE-BRASS TEMPERS

Rolled Temper	Tensile Strength (1000 psi)	Hardness, Rockwell B Thickness (in.)	
		0.020 to 0.036	Over 0.036
Quarter-hard	49 to 59	40 to 61	44 to 65
Half-hard	57 to 67	60 to 74	63 to 77
Three-quarter hard	64 to 74	72 to 79	75 to 82
Hard	71 to 81	79 to 84	81 to 86
Extra-hard	83 to 92	85 to 89	87 to 91
Spring	91 to 100	89 to 92	90 to 93
Extra spring	95 to 104	91 to 94	92 to 95

Rb 88 and above is extra hard. A hardness of Rb 86 means 58,000 PSI yield strength and should handle 52,000 to 54,000 PSI pressures well, if the chamber is correct. This extra pressure capability means higher safe velocities.

The other curve shows what happens when ultrahard brass is annealed by controlled heating. This is apparently used by Winchester-Western on at least some brass. I have some .300 H&H Super-X-primed cases that show the discoloration of heating. They have a .230 head thickness and 83 to 85 Rb hardness. This means around 79,000 PSI ultimate strength and about 62,000 PSI yield strength. In a full-length .33 Wildcat, they perform well.

For some reason, the 7.62 NATO cases seem to be very hard cookies. Probably this is due to the fact that it operates at pressures up to 52,000 CUP in automatic and semiautomatic arms. .30-06 brass is generally of high quality but I checked some .30-06 M2 stuff by a commercial producer at Rb 72-74. This is the sort of brass we don't want in sniper rifle loads. Some brands of 7mm Remington Magnum cases are rather soft. Other brands like DWM and W-W are relatively hard. 7x61 S&H cases I got in 1955 were only 75 Rb, while those of 1972 were 85 Rb.

Head thickness is related to safety and permissible pressures. The Remington 700, 721 and 722 give rather poor case support. They require thick case heads for optimum safety. I saw a .222 Remington case fail in a Remington 722, from too thin a head. The total head thickness was only

When case necks are not annealed, they may split on cases which have been reloaded many times. From left to right are: .30-30, 7mm-06, .33 belted Wildcat and three additional .30-30 cases. Long term storage of loaded cases with severely work-hardened, uncleaned necks can also cause split necks. In some instances, attempting to pull the bullets with a kinetic bullet puller will result in the case breaking off at the shoulder, the neck remaining attached to the bullet.

about .150". The recess in the face of the Remington bolt is also .150" deep. The radius on the end of the chamber of the 722 is about 1/32". The case support was simply inadequate for this ultra-thin case head. No injuries resulted, thanks to the excellent gas-handling ability of the Remington action.

As mentioned earlier, in the March/April 1973 issue of *Handloader*, Bob Hagel and I had an article published on brass hardness, case design and pressure. We concluded that brass hardness. head thickness and internal capacity were the main variables on permissible pressures. Brass hardness was also the largest variable of the three. We also concluded that 80-84 Rb was about the optimum hardness. Below 80 Rb, the strength seems to drop rather rapidly. The graph of yield strength also shows a sharp drop below 80 Rb. Above 84, there seems to be very little gain. Bob and I found quite a variation in hardness in different lots by the same maker. In the .240 Weatherby, it was 8 to 10 points. In Hagel's Model 70 .340 Weatherby, the W-W .375 brass at 82 Rb achieved 2744 fps (feet per second) with the 275 gr. Speer, while a Weatherby case of 77 Rb hardness got only 2616 fps. The Weatherby case head was also thinner than the W-W.

I have been able to correlate brass hardness with primer pocket expansion in .30-06 caliber. A batch of cases fired

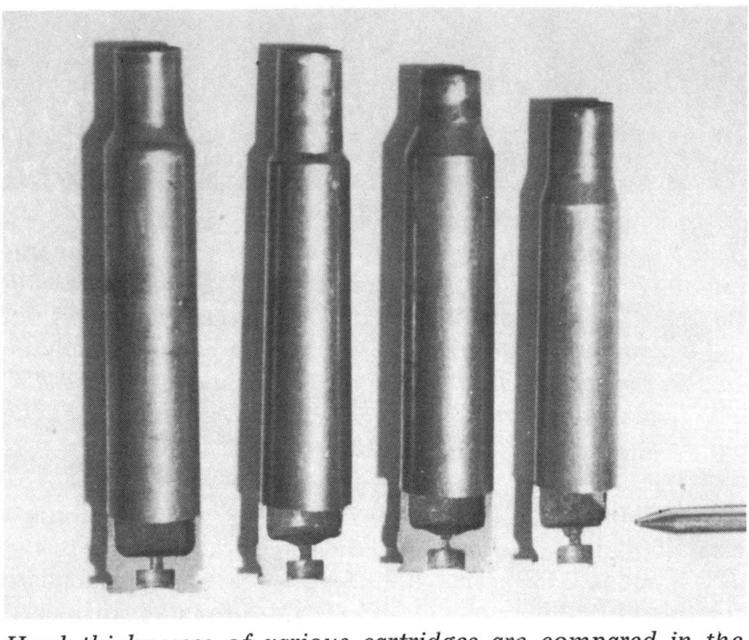

Head thicknesses of various cartridges are compared in the above photo. From left to right are the .300 Weatherby (thin head), .33 Wildcat formed from .300 H&H case (thick head), .300 Winchester Magnum (thick head) and a .33-06 formed from W-W brass (thick head).

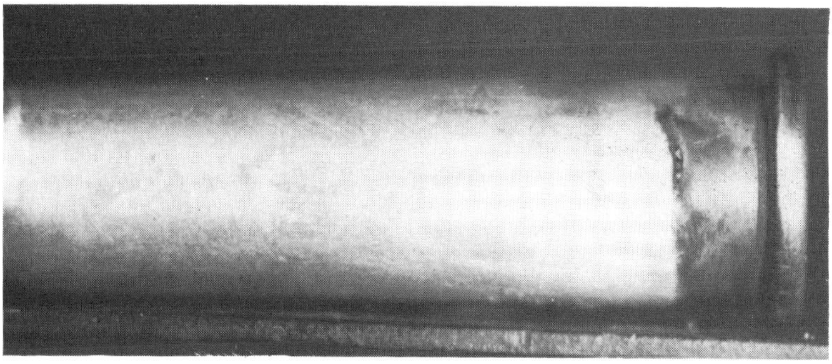

Case inspection is mandatory. This .30-30 case split when fired in a bolt action Mauser bull gun. Even moderate loads are not immune from splitting, even if fired in a rigid bolt action rifle with no head space problems.

with a given high pressure load were sorted into two batches. One batch had expanded primer pockets and the other did not. The average hardness of cases with tight primer pockets was 87.5 Rb, while the cases with loose primer pockets averaged 84.8 Rb. These were arsenal cases and the corresponding yield points are 59,000 PSI and 57,000 PSI. The difference of 2,000 PSI or so is not a large value, but it made the difference between brass stability and instability.

Digressing for a moment, it seems likely that quite accurate pressure determinations could be made in a given rifle by loading successively softer cases until expansion occurred and then using tables based on hardness, case head thickness and chamber clearance.

The significant items from this study appear to be as follows: In belted cases, Winchester-Western cases are usually several points harder than the Remington-Peters line. In .30-06 cases the W-W brand has case heads exceeded only by the hardest arsenal brass, and the W-W are slightly thicker. The commercial R-P .30-06 brass is right up there, however, in the low 80's. In .270 Winchester caliber, no hardness differential existed between R-P and W-W. The DWM-Speer cases in .270 were checked and the uniformity of hardness is amazing. Sako .222 brass is obviously excellent in the few

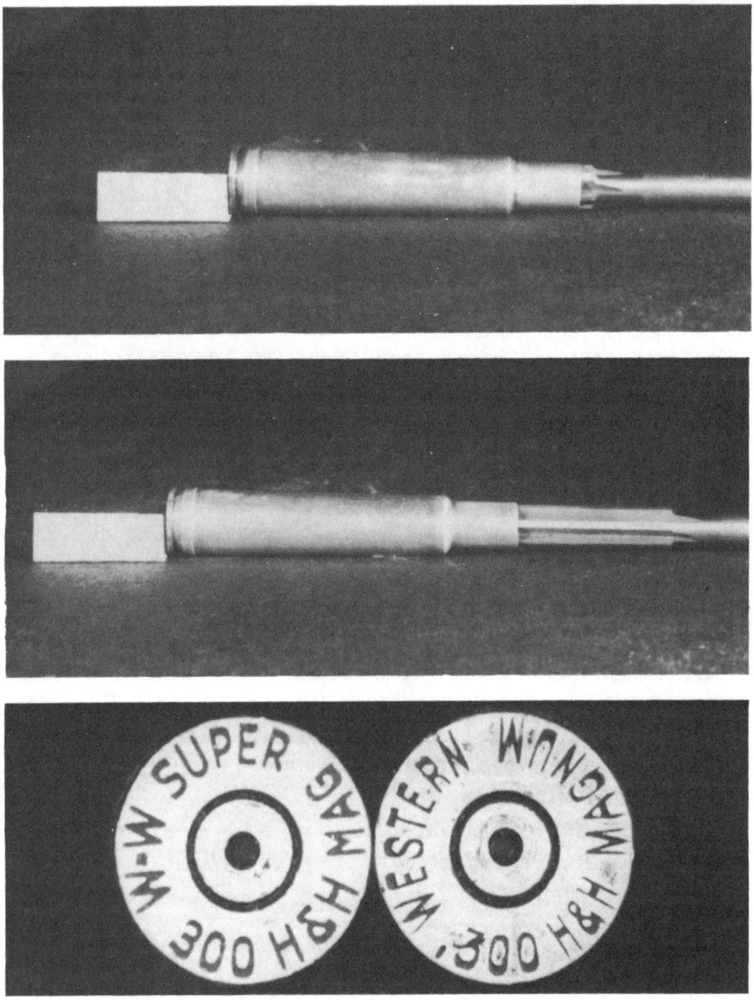

The above photos show the effect of neck thickness on pressure. The W-W Super .300 H&H case weighed the same as the Western .300 H&H. The reamer has a pilot section, then a taper. The W-W Super case neck permitted the pilot section of the reamer to enter while the Western case neck permitted the entire reamer to enter. Thick necks cause more pressure than thinner ones and cause bullets to impact differently, thus raising the dickens with accuracy.

samples checked. The current Federal brass is of extremely high quality with high head hardness and uniformity.

Since velocity is a function of permissible working pressure, the man wanting high performance in a short, light rifle should look hard at cartridge cases based on Uncle's 7.62 NATO case (.308 Winchester). (The idea of a 7mm on that case has long intrigued me.) Now it is a factory round. Similarly, the man who thinks his pet Super Seven is not up to par should check his case hardness and change brands if testing reveals soft cases. There is no point in handicapping a sniper rifle with inferior brass.

Some people condemn World War II arsenal brass. I cannot agree. The hardness values range generally from the low to mid-80's Rb, which is almost identical to many present day .30-06 cases. Among World War II arsenal cases I checked, Den 42's were the most uniform. However, my tests indicate no one should discard any LC43, SL43 or any other arsenal cases of a similar vintage for age alone.

At this time, I wish to express my appreciation to the Penton Publishing Company, the publishers of *Machine Digest*, and Mr. G. C. Strubell, Administrative Director, Metallurgy and Research, Anaconda American Brass Company; Waterbury, Connecticut, for permission to use the graphical data presented in this chapter.

As a postscript, testing in late 1978 revealed our W-W boys were putting out some .375 H&H stuff below 80 Rb. Conversely, .340 Weatherby cases are 86-87 Rb. The only problem is the Weatherby stuff is .183" thick at the head.

The thickness of the cartridge case head has a very large influence on how well it will stand up to repeated high pressure loadings. I like .30-06 size cases to be at least .210" thick at the head and belted stuff .230" or so. I will accept somewhat less thickness if the hardness is there. .215" to .220" is fine in a belted case if the hardness is 82-87 Rb. The weight of the cartridge cases must be uniform, if accuracy is to be any good in a sniper rifle. Think of it this way. The rifle chamber is a fixed volume. Brass weight variation means a corresponding loading density variation; which means pressure and velocity are not going to be uniform. This opens up groups. As an example, a friend of mine had a .30-06 Springfield sporter with a scope sight that gave 2-1/2" groups. These were comparatively large groups at 100 yards. The loads were put up in a motley collection of .30-06 cases, with a weight variation of 12 to 15 grams. Similar loads in uniform cases (+ or - 1 gr.) cut the groups to 1-1/4" to 1-1/2" at 100 yards. This means that cases should be of one make and if possible, all one lot number. Hardness and case head thickness should be checked. The variation in case weight can be surprising in some calibers. In the .308 Winchester, it can be 30 grains, which changes powder capacity by about 4 grains. In 7mm Remington Magnum, weights can vary over 15 grains.

Since Browning brought out their semi-automatic rifle in 7mm Remington Magnum, .300 Winchester Magnum and .338 Winchester, the weights of these cartridge cases have increased. I believe this rifle demanded a heavy case to prevent functioning difficulties. The .338 can be a very bad offender in this. Early cases from the late 1950's will hold a good 2 grains more powder than the current production.

I like to weigh cases, even if they are from the same lot. I once found 10 grains difference in the same box of 8x57 Winchester cases, so don't take anything for granted.

Neck thickness is the next thing. This is an often overlooked item. A thick neck causes more bullet pull, which causes a variation in pressures and velocities. Even when the powder charge is reduced to give similar velocities and pressures with those of a thin necked case, the point of impact will frequently be different. This recently happened to me in evaluating Federal .30-06 vs. Winchester .30-06 cases in a .33 Wildcat rifle. 61 grs. of 4350 in the Winchester case gave similar velocity to 60 grs. in the Federal case, but the point of impact was at least 1" different horizontally between the bullets fired from the two cases. I once lost several points in a 600 yard, 20 shot match due strictly to neck thickness and bullet pull variation. We want to avoid these problems in the sniper rifle.

Inadequate neck clearance can blow a rifle into junk. Don't try to make .243 Winchester or 7mm/.308 cases from .308 or 7.62 NATO cases without making sure neck clearance is adequate. Ream or turn the neck if necessary, to prevent problems. For best accuracy, case necks should be of uniform thickness around the bullet. If the neck is .013" on one side and .016" on the other, it will not help accuracy. Testing will determine just how critical this is in your particular rifle. My experience is that larger calibers are more tolerant than the smaller ones in this respect.

In the .308 Winchester, W-W cases are usually the lightest and will permit heavier charges of a slow burning powder when using 180 to 200 gr. spitzer bullets with corresponding higher velocities than arsenal cases of similar hardness. In the .270 Winchester and the .30-06, Winchester and Federal cases are usually the lightest and also the hardest and would be my first choice in these calibers.

A lot of "experts" seem to think that only a sniper rifle using a military cartridge has any real value. I don't fully agree. A 7mm Remington Magnum is disliked in the silhouette game because it blows holes in the steel targets. The .308 is liked because it doesn't. I want a sniper rifle to have *penetration*. The more the better. Wind drift makes fools of everyone at times and the less a bullet is deflected by wind, the greater the likelihood of a first round hit. It amuses me when I read about trajectory compensator scopes insuring a hit at unknown ranges. A .308 bullet can drift *4 feet* in a 5 mph wind at 1000 yards.

A stiff load in a 7mm or .30 caliber Magnum, using heavier spitzer boattail bullets, will flattern trajectory and diminish wind drift very substantially over what a .308 will do. In addition, the more powerful rifle will penetrate barriers at greater distances than the .308.

Note: Results of the author's tests on brass hardness appear on the following two pages.

BRASS HARDNESS TEST RESULTS

Caliber	Headstamp	Avg. Hardness	Min.	Max.	Comments
50 Mg	Rem-UMC	76 Rb	75	76.5	Uniform
404 Nitro	Kynock	78	76	80	Good
375 H&H	Super spd	87	86	90	Great cases
375 H&H	Western	81	80	82	1939 case
375 H&H	Western	82.5	77	84.5	One was 78Rb & pre-World War II
300 H&H	Super Spd	76	72	79	Soft & thin-headed. Mfg. 1940 or so
300 H&H	Peters	78	76	81	Thick heads
300 H&H	Rem-UMC	76.5	75	78	Thick heads uniform
300 H&H	Rem-Peter	76	73	79	Thick heads uniform
300 H&H	Western (match)	81.5	80	83	Extremely uniform cases, 1940 vintage
300 H&H	Super-X	84	82	85	.230 head thickness
30-06	FA58	82	76	85	Some is soft
30-06	FA59	83.5	79	86	More consistent
30-06	LC62	83	80	86	Good cases
30-06	FA53	81	77	85	No problems many cases used
30-06	TW54	82	80	84	Typical
30-06	LC43	85	80	87	No problems hundreds used
30-06	SL43	85	80	87	Good stuff
30-06	Den 42	84	81	87	Probably best
30-06	Super Spd	84	83	87	Good cases
7.62 (.308)	LC55	86	83	87	Tough stuff
7.62	FC61	82	78	85	Heavy cases
7.62	DA61	91	none		One checked
270 Win.	DWM	80	78	81	Extremely uniform
270	Super-X	86	81	89	One case soft
270	Rem-UMC	87	87	87	One case checked
250 Sav.	Super-X	89	88	91	WOW!
250 Sav.	Rem-Peter	89	88	89	One checked
222	Sako	86	84	87	Great cases
222	Super Spd	87	87	87	One checked
222	Super-X	79.5	78	81	One checked
8x57	Super Spd	82	82	82	One checked
8x57	Rem-UMC	83.5	83	84	One checked
7x57	Super-X	85.5	85	86	One checked
7x57	Western	77	74	76	One checked
7.65	Norma	80	78	80	One checked
7mm Mag.	Rem-Peter	74	72	76	Soft for hot loads
7x61 S&H	Norma	75	74	77	Thick head
375 H&H	Western-SuperX	85	83	87	Made in mid-1950's

Caliber	Headstamp	Avg. Hardness	Min.	Max.	Comments
300 H&H	W-W Super	83	82	85	Made in late 1979 or early 1980
340 Wby	340 Weatherby	87	86	87.5	.183 head thick-ness
340 Wby	340 Weatherby	77	76	78	Old brass
7mm Rem. Mag	Federal	81.5	81	82	Very uniform
300 Win. Mag.	Federal	82.5	82	83	Very uniform
30-06	Federal	89.5	89	90	Excellent brass
270 Win.	Federal	90	90	90	Excellent brass
7x61 S&H	Norma	85	85	85	Excellent brass (recent lot)

CARTRIDGE CASE WEIGHT EXPERIMENT

In order to demonstrate the influence cartridge case weight has on velocity, a test was run using a Winchester Model 88 rifle in .308 Winchester caliber. Data is as follows:

Primer	Remington 9½
Bullet	180 gr. Hornady spire point
Powder	41 gr. of 4064
Heavy cases	LC64 & 65 N.M.; with weights of 189, 185, 188-1/2, 187-1/2 and 185-1/2 gr. primed
Light cases	W-W Super with a weight of 159 gr.
Velocities	2448 fps instrumental with the heavy cases. 2312 fps instrumental with the light cases.
Comment:	When the fired cases were checked for internal capacity, I found 41 grs. in the heavy cases took up as much vertical height as 45 grs. in the light cases. I am certain that it would have been no problem to exceed 2500 fps with the light cases and 44 or 44-1/2 grs. of 4064 with the bullets in these light cases and using the Remington 9½ primer. Actual-

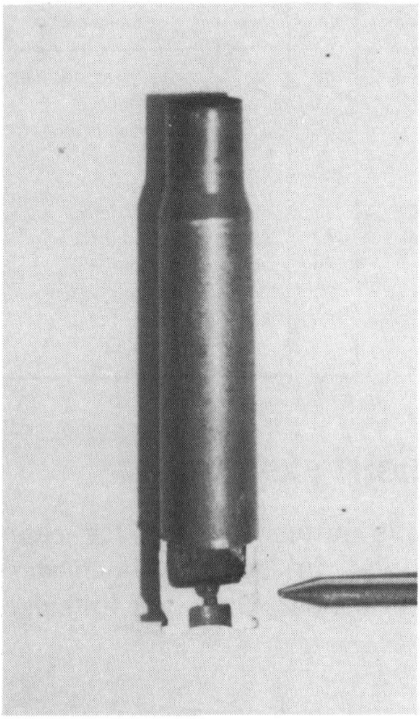

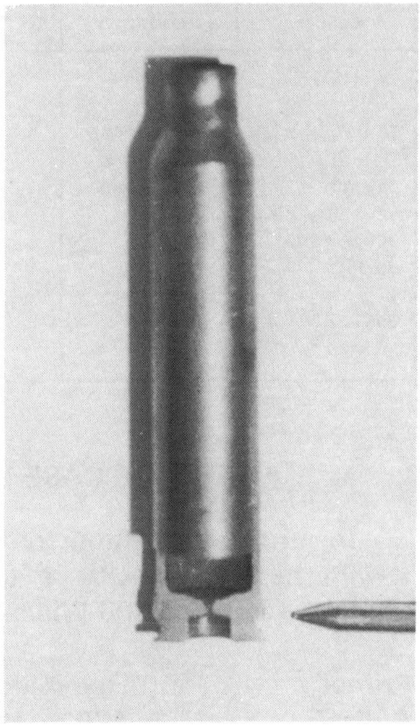

Left: Cross section of rimless case showing where the hardness check is performed on the solid web section. Right: Cross section of a belted case. The hardness check is made on the middle of the belt.

ly, the case capacity of these National Match .308 cases is just about the same as W-W .300 Savage cases. With standard extruded powders and heavy cases in a 22" barrel. the .308 Winchester is more like a .30-40 Krag than a .30-06. The .30-40 was an extremely well regarded cartridge in its day and its ballistics level is still a most useful one for sporting purposes. One thing that these tests brought out, is that the recoil level of these .308 loads is considerably less than of a .30-06.

Case Preparation

This chapter is concerned with safety, reliable functioning, uniform ignition and consistent bullet pull. The last two have a great deal to do with achieving a high level of accuracy.

Let's start off with the base of the cartridge case. On cases of '06 size, a magnum primer may or may not be an asset. Some non-magnum .210 size primers are quite potent, two examples being the Federal No. 210 and the W-W No. 120/8½. On belted cases the size of 7mm Remington Magnum and larger, I find the Federal 215 Magnum primers to be real group shrinkers. This is especially true when using compressed loads of slow burning stick powders. The only difficulty in using these and other Magnum primers is that they are often thicker from front to rear than most other primers and require a somewhat deeper pocket. Some older .375 H&H and 300 H&H cases in my possession have too shallow a primer pocket to accept the Federal 215. This can be corrected by *very careful* use of a .206" diameter spade drill. The amount of brass to be removed is normally only about .005" to .008".

Burrs on the flash hole can be removed by a tool similar to a drill but ground to a much sharper point. Flash hole diameter can be determined by the use of number drills. I prefer to find what is the average size flash hole and then ream all those that are smaller to that size. The few cases that

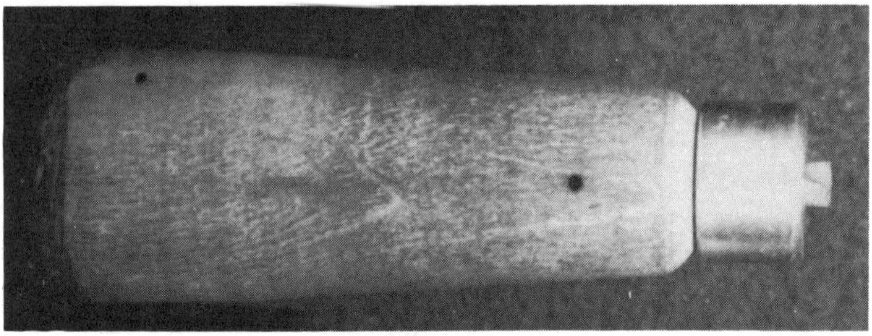

The simple but effective device shown above is used to achieve uniform primer pocket depth suitable for any standard primer, including the Federal 215. A plastic handle would be preferable to the wooden one shown since it would be unaffected by humidity.

have larger flash holes can be set aside for less demanding jobs. Oversized flash holes really can boost pressures. Brown Precision, P.O. Box 270W, Los Molinos, California 96055, has a tool called the "Uniformer" that deburrs and sizes the flash hole in one operation.

When preparing fired cases for reloading, check the cases all over for cracks and splits. Sheet brass can be rolled with an inclusion or defect in the metal. Then when the brass is drawn into a cartridge case, the case could split at the defect on the first firing. Some rifles have rear locking actions that permit the cartridge case to develop a localized stretch point just ahead of the solid rear base of the case. Continued reloading will eventually cause the case to pull apart at the stretched area if firing in a rear-locked rifle is continued.

Even if a front-locked rifle is used, excessive head space caused by improper resizing of the case (or excessive head space in the rifle) can cause stretching and damage. Localized stretching results in a thin section of brass circumferentially around the cartridge case wall, about 1/4" ahead of the solid base of the cartridge case. Usually, it can be visually apparent. A sure test is to use a piece of wire 1/16" in diameter and with a section 1/8" to 3/16" long, bent at right angles to the rest of the wire and given a sharpened point at the bent end.

The wire is inserted into the case clear to its base. Then the pointed end of the wire is pressed against the inside wall of the case and dragged forward slowly. A thin section in the case wall can easily be detected by this method. If you find cases where this defect exists, don't use them for sniper rifle combat loads.

Uniform bullet pull means uniform neck thickness, consistent neck hardness, a non-varying coefficient of friction between the bullet and the inside cartridge case neck, plus total freedom from interference caused by the rifle chamber.

Sniper rifle ammo must be capable of being stored for years with the absolute minimum change in performance that handloading practices can achieve. For this reason, fired cases should have their necks annealed. If cases are not annealed, split necks are almost certain to develop in long term storage. Annealing can be done in various ways. I anneal case necks when I am casting bullets. *A note of caution:* Never anneal more than a neck and shoulder. If the base is annealed, a very dangerous situation is created. I use pliers to hold the base of the case and insert the neck and shoulder into the hot lead for 10 to 15 seconds. I then hold the base of the case under cold running water, until the case is cooled. Do this only with a fired primer in the case. If done without a primer, the molten lead can be trapped in the shoulder area and solidified. *CAUTION:* Never, never use a live primer if you are annealing case necks by this method. Also, wear safety glasses. Extreme care must be taken to prevent heating the base of the case to more than 200° F.

Other methods of annealing case necks involve standing the cases in water to a depth of just below the shoulder area and then using a small torch to heat the area where annealing is desired. This annealing will prevent cartridge case necks splitting from the radial force exerted by the bullet in long term storage. The annealed neck will reduce the force with which the bullet is gripped by the cartridge case and will contribute to accuracy.

Reaming the inside of the case neck will contribute greatly to uniform bullet pull. Cartridge case neck clearance must be sufficient for safety. If a bullet will not freely pass through the neck of a fired case, reaming is mandatory. If all cases are of one make and lot number and neck clearance is ample, then reaming can be omitted. If cartridge cases are necked down from a larger caliber, such as making .25-06 from .30-06 cases, it is a good bet that reaming will be necessary to obtain enough neck clearance for safety and accuracy. Sharp-shouldered, belted cases frequently develop a tight neck on the portion where the neck joins the shoulder after about 3 firings. If the bullet base is seated below the neck/shoulder junction, trouble will arise. The trouble may be merely poor accuracy, but it may be worse. Inside neck reaming will cure this condition. (As a side note, bench rest shooters usually cut away brass on the outside of the case neck to obtain perfect concentricity for the optimum in accuracy. Not being a bench rest shooter, I have simply used inside neck reaming as my standard method where needed for safety and accuracy.)

Neck length must be short enough to eliminate any interference caused by the chamber that would inhibit the bullet leaving the case. I have seen groups go from an inch at 100 yards to 5" or 6" when neck length was allowed to become excessive by only about .010" or so. Many reloading tool manufacturers sell neck trimming tools. They really speed up an otherwise slow, tedious job. High pressure loads will make necks lengthen so check after every firing and trim as needed.

A consistent coefficient of friction between the bullet and the inside of the case neck requires that the inside of the case neck be as clean as when the case was new. Case tumblers, with the right media, seem to do a fairly good job here. I have been testing a Vibra-Tek unit and have found it deficient at cleaning the inside of case necks with the iron oxide-impregnated walnut shells that come with the unit. A more aggressive media would improve this situation, I am sure. Very dirty case necks may need stronger measures. Abrasive cloth or abrasive paper in the 280 to 320 grain size can be torn into strips and then wrapped around a small rod or other round object and used to polish the inside of the case neck. Silicon carbide or aluminum oxide are quite sharp and aggressive in action. However, it is easy to remove too much brass with these man-made abrasives. The less aggressive abrasives, such as garnet, are much milder in action and much easier to use. Some shooters use steel wool to clean the inside of the case neck.

Chemical solutions are sometimes recommended to clean cases but the disposal problems have led me to stay with mechanical means of cleaning cases. Whatever method is used, all the discoloration and firing residue has to come out. That residue will raise bullet pull sharply in the short run and literally cold solder the bullet in after several years' storage. Naturally, accuracy and pressure uniformity become non-existent.

I am a firm believer in chamfering case necks. It allows a flat based bullet, or any other bullet for that matter, to be seated without damage. It also permits a sealant to be used, to moisture-proof the cartridge with minimum problems in application.

Let's discuss cartridge case resizing. First off, we'll forget neck sizing dies when it comes to sniper loads. Regardless of what some experts say, I have always obtained better accuracy when I resized the case full length as compared to neck sizing only. What I believe happens is as follows: No chamber or cartridge case is absolutely concentric with the axis of the bore. Full length sizing allows the bullet to better align itself with the bore since there is no pressure from the body of the case interfering with the alignment of the bullet with the bore. This interference may be present when neck sizing only is done since the cartridge case is a very tight fit in the chamber. Possibly, the super precision bench rest rifle would be an exception to this statement. Cartridge case resizing should be done to small enough dimensions so that rapid fire bolt manipulation is possible on a bolt action rifle. For a semi-automatic rifle, special small base dies are available to permit reliable functioning of the reloaded cartridges. On a standard common caliber such as .270, .30-06 or .308, I belive in sizing the case small enough to achieve good interchangeability with other rifles in that caliber. After resizing, wipe off the cases to remove all traces of sizing lubricant and put them back in the tumbler or Vibra-Tek unit for about 30 minutes or so. When removing them from the device, check the primer pockets for any traces of the tumbling compound. Normally the primer pockets will be clean after tumbling, but if not, use a scraper to clean away the residue. Also make certain none of the tumbling compound has stuck in the flash holes.

The cases are now ready for subsequent operations.

3

Primers And Priming

For a very long interval of time, the quality of primers was something that was taken for granted. Magnum primers, or at least primers that were designated as Magnum, did not exist. The U.S. Ordnance Departement stuck with their corrosive primer mix in .30-06 ammunition until well after World War II. In terms of storage stability, pressure/velocity ratio, and uniform pressures and velocities, it was a most efficient primer. The defect which it had was its corrosive nature in the barrel. The primer residue was a form of salt that reacted with moisture in the air in the rifle barrel and promoted rusting. Since the military had very rigid rules on rifle maintenance, the primer was less of a problem for them than it would be for the average sportsman.

When Federal released their No. 210 large rifle primers to the reloading public in 1952, they were very consistent, efficient primers. It is my understanding that they were originally formulated for a government contract and rather closely simulated the very desirable ignition and storage characteristics of the old FA70 corrosive primer, without the corrosive effect. They had a rounded contour and a brass colored cup.

For a period in the late 1940's, Remington made some large rifle primers designated as No. 9½ that were very hot. They would match most so-called Magnum primers made today, when it came to igniting compressed loads of 4350 and 4831. Winchester large rifle primers formerly came in two varieties. The hot one was called the No. 120 and the milder one, No. 115. The latter was used in cartridges like the .30-30, .25-25 and similar loads. I saw some .300 H&H Magnum match loads made by Western that were being used by the All Army Rifle Team in 1958, using the No. 8½G corrosive mercuric primer. This was a special primer that was once loaded in several large capacity cases by Western prior to World War II. But after 1950 it was only used in special match ammunition.

For years, even bench rest shooters seemed to pay little attention to primers. In the early years, bullet quality was not what it is today and primer variables were masked by the problems with bullets. Also, the .219 Wasp and similar cartridges used powders in the medium burning range. The combination of easily ignited powder in a small case tended to mask primer problems. Perhaps the main thing that delayed the awareness of primers as a real factor in bench rest rifle accuracy was the super heavy barrels that many of the early ones sported. Rifle barrel stiffness is not exactly a linear function of weight, but it is close enough that an educational comparison of the bench rest rifle and the sniper rifle can be made. A bench rest rifle may use a cartridge with 1,200 ft. lbs. of energy from a 10 lb. barrel. The sniper rifle may develop 3,800 ft. lbs. of energy from a 4 lb. barrel. Mathematically, the former has 120 ft. lbs. of energy per pound of barrel weight while the latter has 950 ft. lbs. of energy per pound of barrel weight. So it can be seen that the sniper rifle is almost 8 times as sensitive to ignition variables as the bench rest rifle. In addition, the sniper rifle will likely use from 60 to 75 grains of powder, with large kernels heavily coated to retard the early stage burning rate. So we have the deck stacked against us for uniform ignition in the sniper rifle load. Primer quality and group size have a very definite correlation. About 1960, I was getting groups in a heavy bullet Wildcat rifle that had me puzzled. The rifle's overall group was about 3" vertically by 1" horizontally. However, the vertical dispersion was peculiar. In effect, I had two groups, 2" or more apart, but both the bottom and top groups were quite small. On testing with a chronograph, I found that I was getting two distinct levels of velocity. The lower groups were being made by bullets whose average velocity was 2520 fps. The upper group was being punched out by slugs averaging 2580 fps. The primers were CCI 200's. I tried Winchester 120 primers and got 2595 fps with one ignition level. Federal 210's were also very efficient at igniting the compressed charges of 4831 that I was using.

For a sniper rifle to be worth a nickel for hitting small marks at long range, the vertical dispersion of its group must be small. The primer can have a very strong influence on the size of that vertical dispersion. Consistent velocities, non-varying pressure time curves and uniform peak pressures are a must if a sniper rifle is to obtain the best accuracy it is capable of. I believe that the character of the ignition a primer provides is very important. A long belted case has a very lengthy powder column to ignite. If the powder at the base of the bullet is not ignited at the same instant as that at the rear of the case, it will cause peak pressure variations. Also, the uniformity of the velocity and the time pressure curve will suffer. Some primers tend to raise pressures without having much of an increase in velocity. This is related to the character of ignition that a primer provides. The early Federal 215's acted like that for me, but the late ones do not. As a matter of fact, I am very much impressed with the ability of the recent Federal 215 primers to shrink groups with powder charges in the 65 to 80 grain range, without any unpleasant side effects. Excessive ignition is something the designers of the .22 and 6mm PPC, as well as the .22 and 6mm bench rest cartridges, have been concerned about. All four of these cases use the .175" diameter small rifle primer. The flash hole on the PPC is also quite small.

In a very real sense, we are igniting the powder column from the wrong end. It should be ignited at the front end. I have experimented with front end ignition in a .285 OKH, which is a 7mm/06 Wildcat. It was designed in the late 1930's by the late C. M. O'Neil, Elmer Keith and Don Hopkins. Uniformity of velocity is extremely good and the muzzle flash is reduced. Charlie O'Neil even loaded 5 grains of black powder in the middle of a compressed charge of 4831 in a very large belted cartridge to obtain a more uniform ignition. I have heard of one 6mm experimenter getting 3800 fps with a 100 grain bullet, using front ignition in a large cased 6mm Wildcat. O'Neil and I carried on a lot of correspondence and we had many discussions of ignition during my visits to his home. He patented a front end ignition flash tube under the OKH Duplex label. His belief was that the main virtue of front ignition was to prevent powder jamming in the shoulder area, which raised chamber pressure but did little to accelerate the bullet. Charlie believed that the Duplex system did not offer much advantage in velocity in a large caliber cartridge, where the opportunity for powder jamming was non-existent. The limited tests that I have conducted tend to confirm this. For front ignition to work, it must use a compressed charge and the tube length is quite critical. Incorrectly done, very high pressures can result. Elmer Keith wrote to me of melting case heads completely away and of gas cutting the bolt face of a Magnum Mauser rifle. *Caution:* If any reader wants to work with front end ignition, I assume no responsibility if he comes to grief. I will concede that the potential for very high speeds in small bores with heavy bullets gives it an appeal to the dyed-in-the-wool experimenter, looking for new challenges.

The amount of variation in velocity a primer will give with a large change in air temperature is an important consideration in evaluating primers. I have done some of this the hard way by testing outdoors at 15° F. Winchester 120's still worked well with compressed charges of 4350 in '06 size cases. Velocity is down 30 fps or so from what is obtained at 60° F. Bob Hagel has done some testing on temperature variations with various primers and powders and reported his findings in the May-June 1975 issue of *Handloader*. He found CCI Magnum primers offered no advantages over standard primers in many loads. His findings confirm my own limited testing. The velocity drop with temperatures is not linear. The medium size cases with powders like 4064 are less sensitive to temperature than are the large cases full of slow

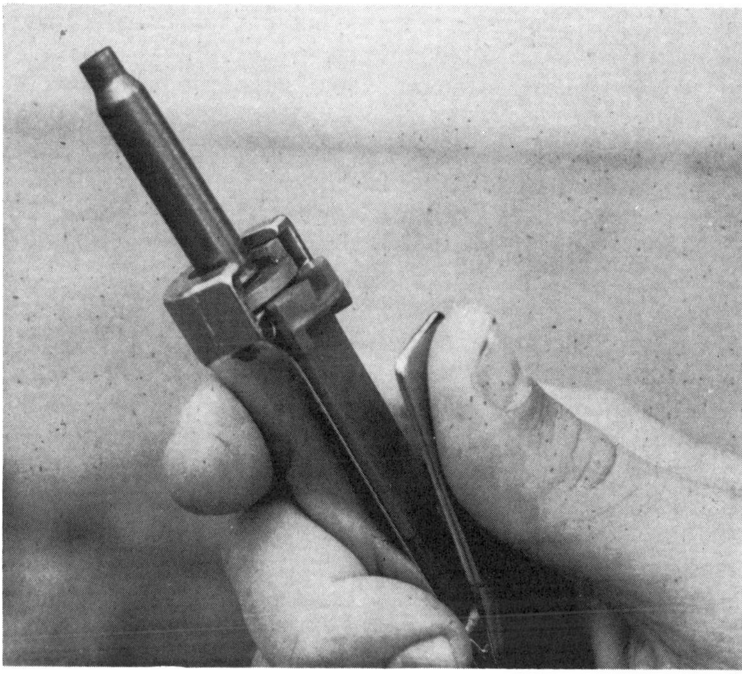

The Lee priming tool allows the reloader to "feel" the primer into the primer pocket of the cartridge case and is a superior method of primer seating since the sense of feel is lost when seating primers in most large reloading presses.

powders. I really believe that the current Federal 210 and 215 primers are very efficient at limiting the velocity change due to temperature changes. This cold weather/hot weather stability is, in my opinion, related to the character of ignition supplied by the primer and its compatibility with the powder being used. Primer dimensional uniformity varies quite a bit. I bought 2000 Winchester 120/8½ primers 5 or 6 years ago and I found that the front to rear dimension varied .004". I even had to sort them with a micrometer for some cases such as W-W .300 H&H. None of my old lots of W-W 120/8½ ever exhibited that amount of dimensional non-uniformity. The cup toughness of the various large rifle primers doesn't appear to vary much, but in the .175" diameter Remington makes a No. 7½ primer that is intended to handle higher pressure than their No. 6½ primer.

My own rating of the potency of large rifle primers would be Federal 215 as No. 1, with W-W 120/8½, Federal 210 and Remington 9½M about even in second place. For the No. 3 spot, I rate the CCI 250 by itself. Remington No. 9½ is quite mild and I place it in fourth position. I have not used CCI 200 primers in 20 years so cannot place them as they are today.

I like to match primer potency with cartridge case size. For 65 to 80 grains, the new Federal 215 is my choice. For 55 to 60 gr., W-W 120/8½, Federal 210 or the Remington No. 9½M would be my pick. In cases of 35-50 gr. capacity, with easy to ignite powders, I'd go for the Remington 9½. With compressed loads or W-W ball powders, a hotter primer is my preference.

The actual priming of the case is best done with a tool sensitive enough so that the handloader can feel the final seating of the primer. Most of the big presses have too much mechanical advantage to allow an adequate sense of feel in a low force operation, such as seating primers. Lee Engineering makes a small one-hand priming tool that is well designed and permits one to "feel" the primer seating. I do one more thing after the primer is home. I turn the case 180° in the shell holder and press the primer one more time. The support of the shell holder is not a symmetrical 360° around the base of the cartridge and this last operation helps compensate for it.

4

Powder Selection For The Sniper Rifle

Smokeless rifle powder can be thought of as a compact form of chemical energy. Powders of the burning rate suitable for a sniper rifle are of three physical shapes: tubular, flake and spherical. Slow-burning flake powders are not available in the U.S.A., although they are available in Europe. I once broke down some F.N. .30-06 military ammo and found 54 grains of a flake powder behind a 150 gr. bullet in a Berdan primed case. Some spherical powders contain flattened grains which control the maximum web thickness. The net result is some tubular powders require less volume for a given weight than others. Powders vary from lot to lot in their density per unit volume also. Chemically speaking, smokeless powders come in two varieties. The type commonly called single base contains nitrocellulose only. The powder commonly called double base has nitroglycerin plus nitrocellulose.

Coating on the outside of the powder influences its initial rate of pressure rise, how difficult the powder is to ignite, the maximum flame temperature and the type and amount of accuracy-destroying gunk it leaves in the barrel upon firing.

So-called cannister powders are supposed to be very consistent from lot to lot. This is not always true. The worst offender I have ever encountered was the now discontinued Norma 205. I have seen lots of N205 whose maximum safe working load varied from 56 to 59 gr. with all other components remaining unchanged and the resulting velocity changed from 2920 to 3060 fps. In fact, the slow lots would push a 180 gr. bullet just about as fast as the fast lots would push a 160 gr. bullet in the particular 7mm Wildcat rifle I was using at the time. Norma 201 has also shown more lot to lot variation than I like. In view of these lot variances, I believe one should purchase sniper rifle powders in 8 pound or larger containers. In addition, very good storage conditions should be maintained for any powder.

The primary requisites in powder evaluation include: accuracy, uniform velocity for minimum vertical at long range, a very favorable velocity/pressure ratio and a muzzle blast/flash combination that is not excessive.

Secondary items in powder evaluation would include: how sensitive the powder is to wide changes in temperature, the storage stability of the powder, its ease of ignition, the tendency of the powder to deposit accuracy-destroying gunk in the barrel, how sensitive the powder is to small variables in bullet pull and primer intensity.

For the sniper rifle, I personally would not attach much importance to a powder's compatibility with powder measure usage, since super accurate ammo for combat use in a sniper rifle should have the charges individually weighed.

In anyone thinks there is an easy method for picking the optimum powder for a certain cartridge with a given bullet, he is in for a rude awakening. There are a number of mechanical variables that influence the pressure/time curve of a powder charge with a fixed weight of bullet in a given cartridge. These break down into barrel variables and bullet variables. The barrel variables include: groove diameter, land diameter, groove depth, land width, the throating of the barrel, the surface finish in the bore, and last but not least, the amount of wear the rifle has had. The bullet variables include: bearing length, bullet diameter, jacket thickness, jacket hardness, core hardness and jacket material.

The primary variable among loading variables is seating

depth, since it influences the bullet's movement until it contacts the lands.

The mechanical sequence of assembling a cartridge is not complex, but the study of interior ballistics is quite complex. If it begins to look like interior ballistics is a major factor in how a powder performs, you are absolutely right.

As an example, take two .30-06 rifles with 24" barrels. Let rifle A have a .309" groove diameter, a long throat and two grooves in the barrel. By way of contrast, let us say rifle B has a .3078" groove diameter, 6 grooves, wide lands and a short throat. If 180 gr. Sierra bullets are going to be used in both, it is a very good bet that rifle A will be much more compatible with DuPont's 4064 or 4320 than rifle B will. The latter rifle will probably do very well with around 55 gr. of 4350. Rifle A has a barrel with the interior dimensions favoring a slowing down of the time/pressure curve. In other words, the rate of pressure rise will be slower and the maximum pressure less with a given powder charge in rifle A than in rifle B. Conversely, rifle B has a barrel that speeds up the pressure rise per unit of time and consequently gives a higher peak pressure reading.

For example, let's take what I consider to be the smallest useful sniper rifle caliber, the .222 Remington, to illustrate some points. A very good load for the .722 Remington back in 1950 and 1951 was 21 gr. of DuPont 4198 and a 50 gr. bullet. It was a nice working load with the cases, primers and bullets of that era. Provided the barrel was good, the load generally produced quite decent varmint accuracy. With a barrel tighter in dimensions and possessing a minimum chamber, plus a short throat, I would lean towards a slower powder such as DuPont 3031 or 4895. Handbook tables of maximum loads vary from book to book. For example, the latest Hornady handbook quotes 20 gr. of 4198 as maximum behind their excellent 50 gr. bullet in the .222 Remington. This is a full 5% less than we used to use 30 years ago. Evidently the components have not remained constant over time or else the particular rifle used by Hornady was very tight in the barrel and chamber. As another example, the DuPont loading charts for 1965 list the 21 gr. 4198 load in

.222 Remington as producing 45,600 C.U.P. with a 50 gr. bullet. The DuPont table for 1975 shows 20.5 gr. of 4198 as maximum in the .222 Remington with a 50 gr. bullet and the pressure is listed as 44,500 C.U.P.

In picking a powder, I prefer one that fills the case to the base of the bullet and is not sensitive to changes in temperature. I also prefer a powder that is not hard to load for accuracy. The spherical powders seem much more sensitive to temperature changes than single based tubular powders, in my experience. I once blew a primer with a load of Winchester 780 ball powder on a hot day, yet that same load had not shown excessive pressure in cold weather. Hodgdon's 450 powder has always seemed more difficult to load for fine accuracy than their H4831 of similar rate. Winchester's newer 748 and 760 powders have given me much better results, but still are more touchy in temperature extremes than I like. The latest Speer handbook cautions against using certain double based powders in extremes of heat or cold.

The new Norma MRP appears to be far superior to the N205 it replaced, but it still has its quirks. In a large case .33 Wildcat it gives speeds closely comparable to H4831 or MR 3100 at a similar pressure level. The only problem is the groups with MRP are about twice the size of those with the other two powders. In a long throated 7mm/06, MRP is very accurate with a splendid velocity/pressure ratio. The 7mm Express Remington is similar to the 7mm/06 and the Norma 7mm Express ammunition is loaded with what appears to be MRP and produces exceptional velocity. In handling the 7mm Express, various writers cited the good results they obtained with MRP. My personal opinion is that MRP is at its best when used in cartridges where the maximum charges do not exceed 70 gr. In cartridges where the maximum charge is higher, I prefer H4831, IMR4831 or MR3100. MRP is very good in .243 Winchester and excellent in .270 Winchester when using the heavier bullets available in those two calibers. At least this has been true in the rifles that I have tested it in.

Norma 204 is a bit slower than the 4350 and considerably more dense. The Norma product is much less flexible and for me at least, it is harder to get fine accuracy than with

4350. The higher density of Norma 204 permits quite heavy charges in cases that are a bit short on powder capacity. Two examples are the .308 Winchester and the 8x57 Mauser.

Frequently full pressure loads using the three types of 4831, MR3100 and sometimes 4350 require the use of a drop tube to get the powder charge in the case so that excessive compression of the powder charge does not result when the bullet is seated. In addition to the drop tube, I find the Vibra-Tek unit is an assist in settling the powder charge. To use the Vibra-Tek unit for this purpose, I use a loading block full of cases charged with the proper charge and use a bullet slightly smaller in diameter than the correct size set into the case neck as a stopper. As an example, for .30-06 cases, I use a 7mm bullet sitting on the powder charge while the loading block is held on the Vibra-Tek unit. Proper placement of the loading block on the Vibra-Tek unit is critical for this operation. I find it best to hold a loading block of cases on that portion of the unit where the cantilever spring is fastened to the base. If positioned too far from this point, the amplitude of vibration will be such that the Vibra-Tek unit will make the powder increase in volume rather than decrease. When properly done, the undersized bullets will rotate in the case neck and in about two minutes some additional compacting of the charge will occur. As stated before, if it is done wrong, there will be no compaction.

When a good hot primer is used, slight compression doesn't seem to hurt accuracy one bit. One advantage of such a loading system is that there is no shifting of the powder in the cartridge case with attendant changes in elevation at long range. In 1958, in the Herrick Team Match at 1000 yards at Camp Perry, I got a nice, juicy, low 4 on one round because I forgot to tap the powder to the heel of the case before chambering the round. When the target is a terrorist armed with a 9mm submachine gun or an AK-47, the shooter with a sniper rifle will have more than enough to occupy his mind without having to remember to tap the powder charge to settle it for a consistent vertical impact.

IMR 4831, in my experience, is a different powder than the current Scottish made powder, called New Manufactured

H4381 or the slower lots of surplus World War II 4831 used by the ton by the handloading fraternity. IMR 4831 probably should have been called IMR 4360. In a .308 Norma with 180 gr. Hornady bullets, IMR performs well. But in a 7mm/06 with 175 gr. Hornady Spire Points having a very long bearing length and far less base area for the powder gasses to impinge upon IMR 4831 builds pressure in one heck of a hurry. I find it faster than the fastest lot of surplus 4831 I ever pulled a trigger on. In short, I consider IMR 4831 as being between 4350 and N.M. H4831 in burning rate. IMR 4831 has one good feature in that it is stable at high pressures. It does not give violent jumps in pressure the way some powders do when you are feeling your way around the upper limit of permissable pressures. MRP and N205 are both much more touchy around the upper limit of the loading spectrum than IMR 4831.

N.M. H4831 is a fine slow powder. One .30-06 load I like is 59 gr. behind a 200 gr. Sierra Boattail Spitzer soft point. Velocity in a 1977 Model 70 with a 22" barrel is 2590 fps and uniform. This load might be excessive in some rifles, so start slow and work up. Mr. Claude Sonday, of Accurate Arms Company, sells a powder that he calls MR 3100. The requirements for MR 3100 to function at its best are: A Magnum primer, a long heavy bullet and a relatively large case for the bore diameter involved. I have been outstandingly successful in only two cartridges with this powder. One is the .308 Norma Magnum with 180 and 190 gr. bullets. The other is a 7mm/06 with 162 and 175 gr. Hornady Spire Point bullets. It was much easier to get a good load with the 175 gr. Hornady than with their 162 gr. bullet. In addition, accuracy is better with this powder when the bullet is close to the lands.

The .30-06 is too small in case capacity for MR 3100 to be at its best even with the 200 gr. Sierra bullet. With standard primers, 56 gr. of MR 3100 gave 2450 fps muzzle velocity from a Model 70 with a 22" barrel. 58 gr. of MR 3100 gave a muzzle blast that was unreal, using standard primers and the 200 gr. Sierra bullet. Merely switching to a Federal 215 Magnum primer stopped the muzzle blast problems and gave a muzzle velocity of 2530 fps with the 200 gr. Sierra Spitzer. Pressure with this load was not a problem in this Model 70.

Some very slow ball powders such as 870 have a reputation for depositing gunk in a barrel, killing accuracy after about 15 shots. 6.5mm Norma match bullets weighing 139 gr. did extremely well in 10 shot groups with this and similar powders when loaded in Wildcats with very large cases. However, when this combination was tried in the Wimbledon at Camp Perry, which requires two sighters plus 20 shots for record, it fell apart. I wouldn't relish cleaning a barrel to restore accuracy when several people were throwing slugs at my position. The long range ballistics of these very specialized rifles such as the 6.5mm on the .300 Weatherby case is extremely good but I would not care to cope with the problems they present in a combat sniper rifle with the powders presently available. The big .30's, such as the .300 Winchester Magnum, .308 Norma Magnum, .300 Weatherby and even the old .300 H&H when loaded with a boattail bullet of high ballistic coefficient to maximum safe velocities are not too far behind the extremely large case 6.5 and 7mm cartridges and have a wider selection of suitable powders.

In assessing a powder, it is often a very good idea to try it with several primers. For some reasons I do not fully understand, certain combinations of powder, bullet abd cartridge are sensitive to primer characteristics. If you have a load that is very close to what you want but is not quite there, try a different primer or two in the same intensity range. It may work wonders.

In most cases, semi-automatic rifles do not permit the wide variation of peak pressure and muzzle pressure a bolt action allows. For them it will be necessary to closely duplicate the powder burning rate of commercial and/or military loads available in a given caliber. As an example, in a .30-06 M1 Garand, IMR 4895 performs better than 4350 when 168 gr. match bullets are used. The high muzzle pressure of 4350 ft. lbs. treats the operating rod of the Garand rather roughly and greatly shortens the sustained accurate functioning of the rifle. When National Match .30-06 loads appeared in 1957, they were loaded with a 173 gr. boattail bullet and 4895 powder. Even commercial semi-automatic rifles were somewhat critical as to powder selection. A semi-automatic rifle that will not function reliably is a hazard to its user in a combat situation.

The various reloading handbooks available today are extremely good sources of material for references. Anyone loading combat loads for sniper rifles should have at least a reloading handbook from Hornady, Sierra, Speer or Nosler. Having all of them is even better. Hodgdon has a loading book for their powders and it contains thousands of loads. DuPont and Hercules have booklets available for their powders. In a sense, the best handloader is the chap with the largest library and the most friends. Capt. Phil Sharpe wrote a great book on handloading in the late 1930's and brought it up to the state of the art as it existed in 1948. A man named Major Earl Naramore wrote a splendid treatise on handloading in the 1950's and it is a classic in its field. The best current magazines for the serious handloader are *Rifle* and *Handloader* from Wolfe Publishing Co. in Prescott, Arizona. The latest handloading book is from Wolfe Publishing Co. and is by Ken Waters. It covers some 14 years of Mr. Waters' writings for *Handloader* and *Rifle*. Mr. Waters is a very observant, perceptive handloader and a fine writer. Another recent book is by my old friend, Bob Hagel. It is particularly noteworthy for its practical comments on exterior ballistics and the bullets available today for hunting purposes.

All books on handloading stress making certain that all powders are labeled correctly. *Gentlemen, they mean it in spades!* So make dead certain you do it!

The age of a powder is supposed to be reflected in its performance. The actual deterioration is a function of time and storage conditions. I store mine in a location where temperature and humidity varies very little over a 12 month period. Damp powder will produce lower velocities than usual by a surprising amount so give considerable thought to storage conditions for your powder and loaded combat ammo.

D-P 5 7.62 NATO

Ideal Powder Characteristics

In a very real sense, none of the presently available rifle powders are very close to being an ideal powder. An ideal powder would have the following characteristics:

1. Extremely long storage life. The W-W ball powders are quite long-lived from all information I have seen.
2. Produce velocities well above those attainable with present day propellants with no increase in the maximum pressure levels.
3. Possess minimum muzzle flash and blast.
4. Have a lower flame temperature than today's powders to attain lengthened barrel life.
5. Be easy to ignite uniformly.
6. Have very little change in velocity and pressure with a wide variation in powder temperature.
7. Be able to produce accuracy equal to or better than obtainable with today's propellants, and be quite easy to load for accuracy.

8. Have a very predictable rise in velocity and pressure with increases or decreases in charge weight in a given cartridge.

9. Be free of the Secondary Explosion Effect that produces violent pressure excursions when a charge of slow burning powder is reduced below a certain level in cartridges such as the .25-06.

10. Possess a high degree of flexibility. By that statement I mean that it would burn cleanly and predictably over a wide pressure range and with a large variation in bullet weight in a given cartridge. The now discontinued Hercules Hi Vel No. 2 was about the best in this regard I have ever used. I believe a lot of handloaders would agree. I have used it with low pressure lead bullet loads, medium pressure loads in the 30,000 psi range and high pressure loads using jacketed match boattail bullets. I have seen it used in cartridges ranging from the R2 Lovell to the .375 H&H Magnum. It always had a good reputation for accuracy.

11. The residue from the powder should not gunk up a barrel with anything that destroys accuracy.

To approach such an ideal powder will undoubtedly mean a triple base, quadruple base or possibly even more complex composition. I believe the tubular shape is the best because of the need to obtain a highly progressive burning characteristic. The flake powders are common in Europe but are relatively unfamiliar to American reloaders. I am not certain how well such a grain shape would be accepted here. From the standpoint of being able to obtain a very complex characteristic, with consequent extended pressure/time and resulting very high velocities, the flake shape of powder grain has considerable merit.

If some enterprising firm or individual ever develops and offers to the handloaders of this country a powder incorporating as close an approach as possible to the theoretical ideal, it would render presently available propellants obsolete in very short order. The DuPont offerings are basically 40 to 45 years old in the state of the art of powder making.

It would be nice to be able to get 3000 ft. lbs. of energy routinely with a .308 Winchester using heavy military cases and a variety of bullets. Similarly, it would be even nicer to move a 180 gr. bullet at 3300 or 3400 fps muzzle velocity from a .300 Magnum or .308 Norma Magnum using barrels of normal length. Such a powder would boost performance of many other cartridges. For the sniper rifle, such an ideal powder would greatly improve steel penetration and destructive potential on vehicles when teamed up with bullets of suitable construction.

6

Bullets And Barriers - Trees And Logs

Trees and logs, natural obstacles to a bullet's path, have been the subjects of some study by or own Ordnance Department in decades past. Some manufacturers of bullets that are intended for dangerous game in Africa routinely test their products on trees for structural integrity and the ability to hold a course. When Hornady introduced their .375 caliber, 300 gr. Full Metal Patch round-nosed bullet in the mid 50's, they shot it through cottonwood trees well over 3 feet in diameter from a .378 Weatherby Magnum. Randy Brooks, who makes the Barnes line of custom made bullets, has had similar penetration capability from Full Metal Jacket bullets of his own manufacture. These bullets are of a round-nosed design and lose velocity rather quickly as compared to a spitzer. However, they do have great ability to hold a course in going through a tree or log. They do not expand so they are relatively efficient in terms of penetrating ability per ft. lb. of energy in wood.

Expanding bullets are the poorest of all in getting through wood. They open very quickly to a large frontal area and expend all their energy in only a few inches of penetration. In many cases, the impact with wood will literally grind the bullet into confetti. This really is not too surprising since they are engineered to expand quickly on impact with game.

The Full Metal Patch bullets, such as the .30 caliber, 173 gr. government boattailed match bullet, does not hold course well in wood. Neither does the .30 caliber, 150 gr. M2 bullet. I attribute that to the very short cylindrical bearing portion of the bullet and the comparatively long pointed portion of these bullets. They are quite efficient at slipping through the air without losing much speed but are not reliable in trees and logs.

If a law enforcement officer or a citizen is preparing his rifles for dealing with someone throwing slugs from behind a 15" tree and wants to be effective in retaliation, he will need a lot of penetration from his rifle.

The following is some real world experience with shooting into wood.

TEST A1

Rifle:	Winchester Model 70 with a 26" barrel, .300 Weatherby Magnum rechambered from .300 H&H.
Bullet:	Nosler 180 gr.
Powder:	H4831 – 82 grs.
Primer:	Not recorded.
Cases:	Not recorded.
Velocity:	2985 fps instrumental at 17 ft.
Test results (at 15 ft.):	On Maple stove wood cut from green trees only a few weeks before, this load penetrated only about 7-1/2". The bullets lived up to their advertised claims and retained about 2/3 of their weight.

TEST A2

Rifle:	Same as in test A1.

Bullet: Hornady 190 gr. Match Bullet.

Velocity: 2850-2900 fps muzzle velocity.

Test results (at 15 ft.): On the Maple stove wood used in Test A1, this load would penetrate about 6-1/2". The bullet did not expand immediately but when it did, it was rather thoroughly shattered.

TEST B1

Rifle: .33 Wildcat on full length H&H belted case. 26" barrel.

Bullet: Barnes 250 gr., .032 Jacket Soft Point.

Velocity: 2890 fps muzzle velocity.

Test results (at 15 ft.) This bullet penetrated 5-1/2" into the block of Maple and when recovered weighed only 91 grs. The remainder of the bullet was simply ground into confetti.

TEST B2

Rifle: Same as used in test B1.

Bullet: Special Barnes 250 gr. Spitzer soft point, .049 jacket, thinned to approximately .025" at the point.

Velocity: 2900 fps, approximate.

Test results (at 15 ft.) This bullet went slightly over 6-1/2" into the block of Maple, but when recovered, weighed 190 grs. Wood destruction was impressive, but penetration was entirely inadequate to shoot through a sizeable tree.

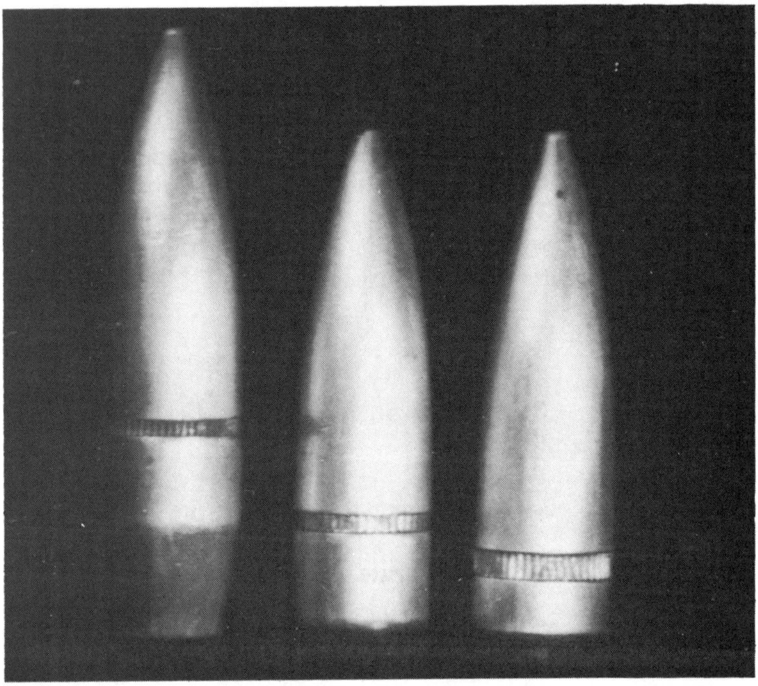

Top: Side views of three bullets having considerable variation in jacket thickness. From left to right are the 173 gr. M1 .30 caliber Boattail made in the late 1920's or early 1930's; 150 gr. M2 bullet pulled from a 1957 FN .30-06 military round; 154 gr. bullet pulled from Winchester-made 8x57 military round. This latter bullet has a considerably heavier jacket than the two .30 calibers. Its Spitzer design gives good remaining energy at long range but rather poor penetration in trees. Bottom: Base views of the same three bullets.

The variation in jacket thickness of the three bullets on the opposite page is clearly shown in this sectional view. The 8x57 Mauser on the right has the thickest jacket. This round did rather well on the car door penetration tests which are covered in a later chapter.

TEST C1

Rifle: Post-1976 Model 70 in .30-06.

Cases: Arsenal.

Primers: Federal 210.

Powder: 4350 – 56 grs.

Bullet: 173 gr. Gov't. FMJ boattail.

Velocity: Not recorded.

Test results (at 30 yds.) On a 12" section of living Cherry tree, this bullet went on through but when it struck a 1/16" section of steel plate leaning against the tree on the far side, it was traveling sideways and left a dent 1/2" deep x 1" wide x 1-1/2" long. Obviously, for the bullet to fail to go through the plate, it would have to be traveling at a very slow rate, probably 500 fps or so.

TEST D1

Rifle: .308 Norma Magnum Bullgun with 28" barrel.

Bullet: 165 gr. armor piercing, pulled from .30-06 military.

Powder: MR 3100 – 72 gr.

Primer: Federal 215.

Cases: Norma.

Velocity: Not checked. 3100 fps estimated.

Test results (at 30 yds.): When shot into a living Birch tree, 14" to 15" in cross section, none of the bullets penetrated. The higher velocity of this rifle gave a slight improvement as compared to a .30-06, but accuracy was still very mediocre. I believe that these bullets tumbled after a few inches of penetration.

TEST D2

Rifle: Same as for test D1.

Load data: Same as D1 except for bullets, which were

	172 gr. boattail and 173 gr. Military Match types.
Test results (at 30 yds.):	On the same tree as used in test D1, none of the bullets penetrated. I believe that these bullets lost stability and tumbled in the tree after a few inches of penetration.

TEST E1

Rifle:	.256 Newton with 24″ rustless steel barrel.
Bullet:	139 gr. Norma Full Metal Jacket Boattail (mild steel jacket).
Powder:	H4831 – 51 gr. (Hot lot.)
Cases:	LC69 reformed, reamed and trimmed.
Primers:	W-W 120/8½.
Velocity:	2860-2900 fps.
Test results:	On the 14″-15″ diameter Birch tree used in tests D1 and D2, none of the bullets fired using this load penetrated. The bullet stability of this 8″ twist rifle is excellent. The problem is that the bullet shape lacks bearing length to prevent tumbling and course change.

TEST F1

Rifle:	8x57 Mauser with 23-5/8″ barrel.
Bullet:	154 gr. Full Metal Jacket Flat Base Spitzer with an exceptionally thick jacket.
Powder:	53 gr. ball powder in W-W loads, 49 gr. same powder in handloads.

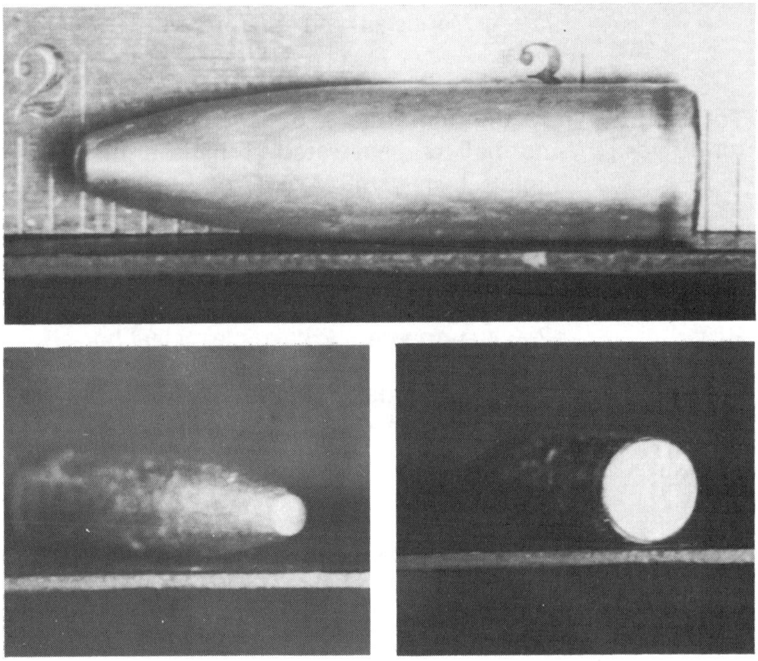

Top: Side view of Barnes experimental 190 gr. .30 caliber bullet with closed semi-Spitzer point and open base. This bullet would penetrate over 14" of a live Birch tree when fired from a .308 Norma Magnum. The long parallel sides gave it good stability in wood, but the open base made it rather fragile on harder targets. Bottom: Nose and base views of the Barnes bullet.

Cases:	W-W in factory loads, Herters in handloads.
Velocity:	Not checked.
Test results (at 30 yds.):	None of the shots fired would penetrate the 14"-15" section of living Birch tree.

TEST G1

Rifle:	Same as on test A2.

Cases:	W-W .300 Magnum reformed and trimmed.
Primers:	Federal 215.
Powder:	MR 3100 – 70 grs.
Bullet:	190 gr. Barnes Semi-Spitzer Solid Point open base, experimental bullet.
Velocity:	Not checked.
Test results (at 30 yds.):	This bullet went through 14+ inches of Birch tree and showed quite good stability when it went through the 1/16" steel plate.
Comment:	This series of tests rather clearly revealed that even powerful, high velocity rifles with tough, expanding bullets are hopeless at shooting through a sizeable tree. The 173 gr. Gov't. boattail FMJ Spitzer was unstable but did a whole lot better than any expanding bullet even when the latter was fired from considerably more powerful rifles. About 1961, while a guest of C. M. O'Neil at his home near Alberton, Montana, he and I shot some quite powerful rifles with expanding bullets into 15" or 16" diameter blocks of stove wood. None of the bullets went more than halfway through the blocks. These rifles had on numerous occasions given one shot kills on elk for Mr. O'Neil and his son.

D–P 7 7.62 NATO

Bullets And Barriers - Automotive Engines

Halting an automobile without endangering the occupants is sometimes necessary for the law enforcement officer.

Damage to the power system can result in immediate halting of the car or a delayed stoppage, depending on where the damage to the car's engine and its supporting systems occurs.

Immediate halting of the engine would result from major damage to the carburetor, the distributor, certain components of the electrical system and the shattering of a cylinder wall by a very severe bullet impact.

Delayed halting of the car would result from damage to the water pump, the radiator and the engine water jacket in the case of a water-cooled engine.

The radiator of any car is very fragile and just about any centerfire rifle possesses enough energy to cause a radiator to leak water severely. The water pump is a very small target and is made of quite heavy gauge cast iron in some cases, of aluminum castings in others. The aluminum is easy to damage but the cast iron can be surprisingly tough material.

Hitting a carburetor or a distributor is a very iffy thing because the location of these parts varies on individual cars. Some V-8 cars have the distributor on the top front of the engine and some on the top rear of the engine. The carburetor is usually top center on most V-8 engines and on the side on an in-line-cylinder job.

By far the largest target in the engine compartment is the engine block itself. My tests have shown that it can be astoundingly resistant to bullet impact, depending on just where the bullet lands. Until a person has fired a fair number of shots at a car's engine block that is hidden from view by a steel plate or some type of screen, that person has no idea of the number of ribs, bosses, locating pads and other areas that contain thick cast iron there are in the engine area.

Car engines made in the 1960's and early 1970's were constructed for durability and long life. The engine block I have been using for tests is GM number 3892657 and it is made of really high grade alloy cast iron. When the reader sees how little damage a .375 H&H Magnum 270 gr. Hornady soft point will do at 25 yards when the angle of impact was just right to be wrong, he will be amazed. This V-8 engine block is one tough baby. Friends who are expert mechanics tell me that recent auto engines are of lighter construction but are still made of an excellent alloy cast iron. Some engines are aluminum and are more easily damaged by a rifle bullet.

The particular engine block used in my tests was probably about as heavy an engine block as would normally be encountered. I carried it around a half-dozen times from my car trunk up a 25% slope some 125 feet from my car, then back to the car after each test. Since I did it alone and unaided, I was glad I had joined a gym earlier that winter!

To my mind, halting a car quickly can be accomplished by a large number of small projectiles in a short space of time, or a few shots featuring extremely tough bullets from very powerful rifles. Evidently, many law enforcement agencies have adopted the hail of small bullets route as Mr. Joyce Hornady, the founder of the Hornady firm, told me that his firm sells the 55 gr. .224" diameter Full Metal Patch Spitzer Boattail bullet by the millions to police departments around the country. No question about it, there is a far greater chance of hitting a distributor or carburetor if 20 shots are fired than if only 2 are fired. Also, those .223 FMJ bullets will penetrate 1/16" steel and still go through an engine water jacket where it is thinnest at 200 yards range. On multiple impact barriers, these small 55 gr. bullets from a .223 possess only marginal stability and cannot be relied upon to maintain a consistent course after impacting such barriers. Also, the car door tests indicate that these small bullets are relatively fragile. I am not surprised at the poor showing of the .223 in penetrating brush in Vietnam.

TEST 1

Rifle:	.375 H&H Magnum, Remington Model 700 with 26" barrel.
Cases:	W-W .375 H&H.
Primers:	CCI 250.
Powder:	4064 – 70-1/2 gr.
Bullet:	270 gr. Hornady Spire Point.
Velocity:	Not tested.
Engine block test:	The scopes that were with the rifle were unusable since the reticle adjustments had been put out of action either by recoil or travel. The result was that this test was done at 25 yards with iron sights. This bullet hit a large boss where it joins a rib and blasted away some of the cast iron boss but created only minor cracks in the water jacket. A 1/16" steel plate had been set up in front of the engine about 12" away. This simulated the real world when attempting to stop a car. This shot amazed everyone who examined the results. Evidently, this hard, wear-resistant alloy cast iron engine block was a great deal tougher than folklore had indicated if hit at

a reinforced point. The impact energy of this load was certainly over 4000 ft. lbs., yet the bullet simply shattered on the reinforced boss. This soft point bullet might be a good game bullet but it did not show much tenacity on this material.

TEST 2

Rifle:	.270 Winchester Custom Mauser with 26" Varmint weight barrel having a 12" twist.
Cases:	W-W .270.
Primers:	W-W 120/8½.
Powder:	4064 – 49-1/2 gr.
Bullet:	130 gr. Sierra Boattail.
Velocity:	3135-3160 fps.
Engine block test:	This was performed at 200 yards with a 1/16" steel plate in front of the engine block. The shot hit a circular boss on the water jacket, where a soft plug is located. The bullet blasted through the full cross section of the boss, some 3/8"x3/8" in size, and knocked the plug out completely. The bullet did no damage to the cylinder itself.
Comment:	This load had too fragile a bullet to damage the cylinder wall after what it did to the soft plug boss. The same hit on an operating engine would have emptied the cooling system and caused the engine to seize in 2 or 3 minutes.

TEST 3

Rifle:	.222 Remington, Savage Model 340 with 22" barrel.

Bullet: 50 gr. Hornady SX.

Powder: 4198 – 19-1/2 gr.

Velocity: Not checked.

Engine block test at 200 yards: The bullet went through the 1/16″ plate, then hit the water jacket where the metal was only about 1/8″ thick and penetrated through. The shot placement was between 2 cylinders and no damage whatever resulted to the cylinders.

Comment: It was a distinct surprise that this supposedly fragile bullet did not blow apart on the 1/16″ plate and spray the engine with bullet fragments. The hole in the engine block water jacket was small, perhaps the size of a .45 bullet or slightly less. It would have emptied the cooling system on an operating engine in a couple of minutes.

TEST 4

Rifle: .308 Winchester in Winchester Model 88.

Cases: LC National Match.

Primers: Remington 9½.

Bullet: 172 gr. M1 Full Metal Jacket Boattail.

Powder: 4064 – 41 gr.

Velocity: Not checked.

Engine block test at 200 yards: This bullet hit just above a boss and went through the water jacket but it passed between 2 cylinders and produced no additional damage.

Comment: The bullet went through the water jacket where it was at least 1/4" or more thick. The cannelured bullet merely sprayed lead on the cylinder wall itself, 1-1/2" beyond the water jacket. The hole in the water jacket was surprisingly small, about the size of the base of a .30-06 cartridge. It would have disrupted the cooling system in a couple of minutes but certainly would not have been an instant stopper.

TEST 5

Rifle: .223 Remington in a Sako-Colt with 24" barrel.

Ammo: South Korean load duplicating the government load with a 55 gr. FMJ Spitzer.

Engine block test at 200 yards: One shot passed through the 1/16" steel plate in front of the block, then penetrated the water jacket but did no damage whatever to the cylinder directly behind the water jacket. One shot went through the 1/16" plate, then hit the inside wall of a cylinder and made a crater approximately 3/64" deep in the cylinder wall. If a cylinder head had been bolted on the engine block, this type of hit would obviously have been impossible. The impact was at a 45° angle to the bullet's path and the lack of directional stability certainly was revealed on this shot.

TEST 6

Rifle: .30-06 in post-1976 Winchester Model 70.

Cases: Arsenal.

Primers:	Federal 210.
Powder:	4350 – 56 gr.
Bullet:	Hornady 180 gr. Spire Point.
Velocity:	Not recorded.
Engine block test at 200 yards:	The first shot went through the 1/16" steel plate and hit an engine mounting boss in line with the No. 2 main bearing on the side of the engine. The boss lost considerable metal but no damage to the water jacket resulted. The next shot hit where the water jacket ends at the top of the sidewall, about 1" down from the head mounting surface next to a boss and rib. This bullet broke through the water jacket but went no further.
Comment:	This bullet sprayed the cylinder with shattered fragments but did no damage of any significance to the cylinder. The cooling system would have been out of order in a couple of minutes but nothing of a quick disabling nature resulted from the second shot and the first would not have disrupted the cooling system.

TEST 7

Rifle:	.308 Norma Magnum Bullgun with 28" barrel.
Cases:	Norma.
Primers:	Federal 215.
Powder:	MR 3100 – 72 gr.
Bullet:	180 gr. Hornady Spire Point.

Velocity:	3100 fps (approximate).
Engine block test at 200 yards:	The first shot hit very high on the sidewall next to where the cylinder head bolts on. The bullet broke through the water jacket and into a tapped hole but did no damage to the cylinder itself. The second shot went through the water jacket and between two cylinders, with no additional damage. The third shot hit the water jacket directly over the No. 2 cylinder, the bullet passing through the water jacket but doing no visible strutural damage to the cylinder wall. All shots first went through a 1/16" steel plate before striking the engine block.
Comment:	Three shots from a powerful .30 caliber Magnum rifle did not produce disabling damage to this engine block.

TEST 8

Rifle:	.33 caliber Wildcat on full length H&H belted case.
Cases:	Western .300 Magnum Match.
Primers:	Federal 215.
Powder:	H4831 (Lot 1969) – 79 gr.
Bullet:	250 gr. Barnes Spitzer with .032" jacket.
Velocity:	2875 fps, approximate.
Engine block test at 200 yards:	This bullet blasted through a heavy boss and rib on the water jacket wall and created a hole over 1-1/2" long by an irregular 1" wide. The bullet hit between two cylinders and did no damage to either cylinder.

Comment: This hit would have drained the cooling system in a few seconds but did absolutely nothing in the way of disabling damage to the engine that would have stopped a car instantly. The structure on this bullet had to be reasonably good to destroy the heavy rib and boss while crashing through the water jacket, but it was totally inadequate to damage the cylinders afterward.

TEST 9

Rifle: .375 H&H Magnum, Remington Model 700 with 26" barrel.

Cases: W-W .375 Magnum.

Primers: CCI 250.

Powder: 4064.

Bullet: 300 gr. Hornady, round-nose Full Metal Jacket.

Velocity: Not checked.

Engine block test at 25 yards: As on Test No. 1, this test was done at 25 yards. This super tough bullet was right in its element as it passed through the 1/16" plate, then the water jacket and knocked out a section of the cylinder wall at least 1-1/2" in diameter.

Comment: This was the first bullet with a tough enough structure to knock out a car engine with one shot. This rifle and load combination has been used in Africa to collect Cape buffalo and elephant. This shot revealed the cylinder wall to be brittle as the portion of the hole next to the piston was 3/8" larger in diameter than where the bullet impacted from the water jacket side.

TEST 10

Rifle:	.308 Norma Magnum Bullgun with 28" barrel.
Cases:	Norma.
Primers:	Federal 215.
Powder:	MR 3100.
Bullets:	165 gr. armor piercing from .30-06 military ammo.
Velocity:	Not checked, but estimated at 3100 fps.
Engine block test at 200 yards:	The first shot passed through the 1/16" plate, then the water jacket and went between two cylinders. Unlike the performance of lead cored bullets, this one knocked out a 5/8"x 1-1/8" hole in one cylinder and a 1/2"x7/8" hole in the other cylinder. The cylinder walls were quite thin at this point (only about 1/10" or so). This shot would very likely have stopped an operating engine, or at the very least slowed it down considerably. The impact of this bullet did not show tipping on the 1/16" plate.

The second shot showed tipping on the 1/16" plate, then went through the water jacket. The bullet deflected and came through a cylinder wall, leaving a small 1/4"x1/2" hole, and grooved the inside of the cylinder wall on the far side of the cylinder.

The third shot showed considerable tipping on the 1/16" plate and was off point of aim considerably. It deflected after the first impact and struck the water jacket at an angle and deflected out a previous hole in the cylinder wall.

Comment: This batch of armor piercing ammo had both good and bad bullets. The good ones showed no tipping on the target and were accurate. The bad ones showed tipping and poor accuracy. The very hard core in these bullets is very destructive on engines when dynamic balance of the bullet is good. When dynamic balance is poor, the bullets are not reliable in action.

8

Bullets And Barriers - Military Style Bullets On Steel Plates & Car Doors

TEST A

Rifle:	.256 Newton, with 24" rustless steel barrel, 8" twist.
Cases:	LC69 reformed, reamed and trimmed.
Primers:	W-W 120/8½.
Powder:	H4831 – 51 gr.
Bullet:	Norma 139 gr. Match Boattail with a steel jacket.
Velocity:	2860-2900 fps.
1/2" steel plate results:	Not applicable.
Car door A results:	Two shots were fired and as luck would have it, they landed practically touching on the

outer panel. Exit holes on the reinforcing panel were 7/16″ diameter. The window guide was hit on one edge. The tear on the inner panel was 1-1/2″ long. The two shots landed practically touching on the 3/8″ backing plate, creating an oblong crater 1/8″ deep and a bulge in the plate 1/16″ high.

Comment: This was amazingly good performance from a bullet this small. Evidently, the jacket strength is much higher than with domestic copper alloy bullet jackets. A lot of powder was burned in other much larger cartridges trying to equal this performance. The directional stability of this bullet in going through the car door was extremely good.

TEST B

Rifle: .308 Norma Magnum Bullgun with 28″ barrel.

Cases: Norma.

Primers: Federal 215.

Powder: MR 3100 – 72 gr.

Bullets: 173 gr. National Match Boattail (metal case).

Velocity: Not recorded.

1/2″ steel plate results: Complete penetration on every shot at 200 yards. The entrance holes are quite a bit smaller than those made by 180 gr. soft point bullets and this load.

Car door B results: The first shot at 200 yards went through the outer door panel, then through the reinforcing panel at the bottom where the two plates are

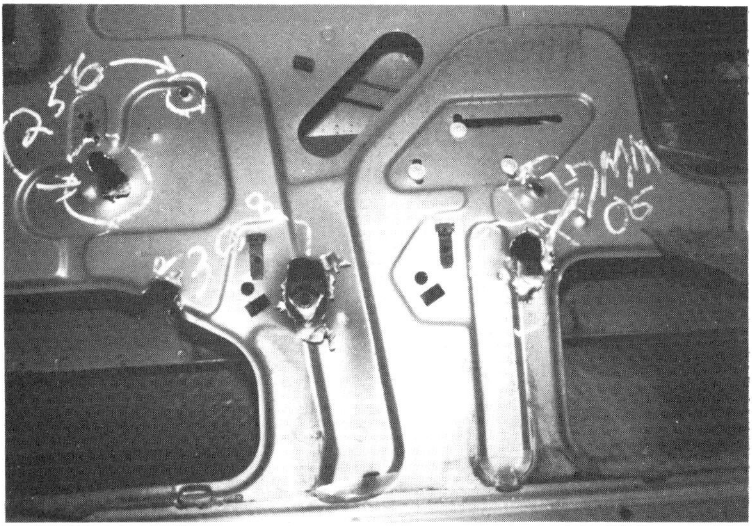

Comparison of the holes in the inner panel of car door A. The holes were made by the .256 Newton, .308 Norma Magnum and the 7mm/06 using soft point and hollow point commercial bullets. The bottom edge of the two ply reinforcing panel is visible in the openings in the inner panels. The upper shot from the .256 Newton missed the reinforcing panel, expanding very little and making only a small hole in the inner door panel.

welded together into a single thickness. The hole in the inner panel was 1-3/8" across with small tears around the hole where shrapnel from the bullet went through. The impact on the backing plate created a crater 1/8" deep and 5/8"x3/4" in size.

The second shot penetrated the outer panel, both sides of the reinforcing panel and left a 1" hole in the inner panel amd a 5/8"x3/4" crater in the steel backing plate 1/8" deep.

Comment: This was a fair performance from a copper alloy jacketed bullet. A much thicker jacket

Holes in car door A made by a .308 Norma Magnum (left) and a 7mm/06 (right). The 162 gr. Hornady 7mm Match bullet has too thin a jacket for this task. The 180 gr. soft point in the .308 Norma came through as shrapnel and lead spray.

would improve its performance very dramatically. The muzzle energy of this load was about 50% greater than the one in Test A, yet the impact on the steel backing plate was very little if any deeper.

TEST C

Rifle: .30-06 Winchester Model 70 with 22" barrel.

Cases: Arsenal.

Primers: Federal 210.

Powder: 4350 — 57 gr. (This charge proved excessive and should be reduced 1 or 2 grains.)

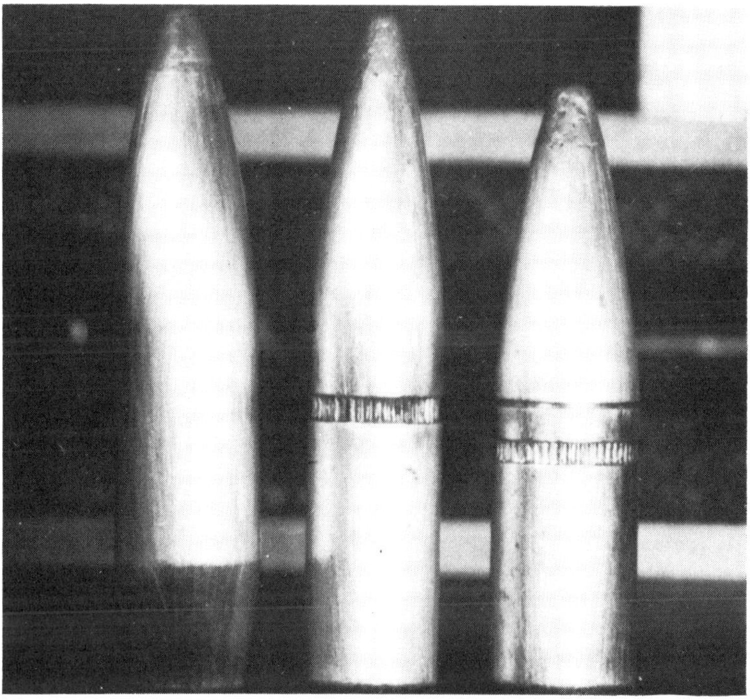

Left to right: 200 gr. .30 caliber Sierra Boattail; 175 gr. .284 Hornady Spire Point; 180 gr. .30 caliber Hornady Spire Point. The structural fragility of the Sierra bullet made it less effective than the Hornady bullets on car doors.

Bullet:	172 gr. metal case boattail with cannelure. Purchased from DCM about 1930.
Velocity:	2730 fps, approximate.
1/2" steel plate results:	This load created craters over 7/16" deep at 200 yards and a bulge on the back of the plate 1/4" high.
Car door B results:	One shot went through the outer panel, then both sides of the reinforcing panel,

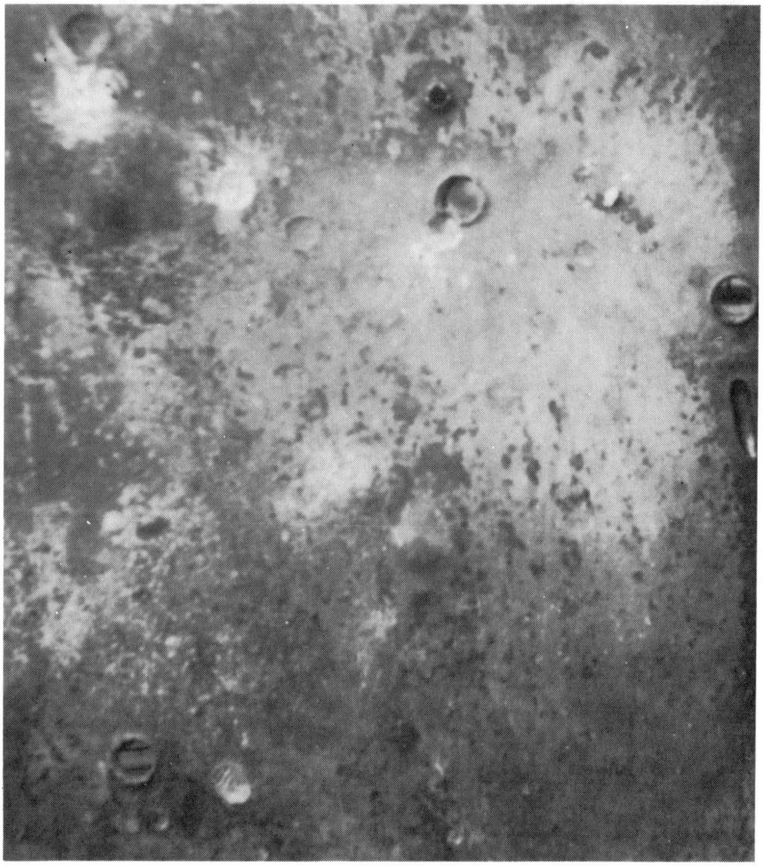

The various lead spray patterns on this plate are what you get when you try to shoot through a car door at the thickest and strongest section with lead core bullets.

leaving a .45 caliber exit hole. The hole in the inner panel was 1-3/8"x1-3/4" and the bullet merely made a spray of lead on the steel plate. One shot penetrated the outer panel, both sides of the reinforcing panel where it also left a .45 caliber exit hole, then blasted a 1-1/4" hole in the inner panel. The bullet fragmented. Its impact on the steel plate was 1/16" deep x 1/2" diameter. Lead spray was found on the plate from this shot around the shallow crater impact area.

Comment: It appeared that the cannelure in this bullet had a very substantial weakening effect on its ability to withstand severe impact stresses.

The 3/8" steel plate used as a backing at 200 yards is shown above. The crater in the plate was made with a 180 gr. Hornady Spire Point fired from a .308 Winchester at about 2450 fps. The turned brass bullet is shown for size comparison only. The two holes were made by a .22-250 using a 60 gr. Hornady bullet traveling at nearly 3600 fps. Note that the holes are quite a bit larger than a .22 caliber bullet. A loaded .22-250 round is shown for comparison. High velocity is essential for a lead core bullet to have much penetration in steel.

TEST D

Same rifle and components as in Test C except the bullets were pulled from .308 National Match ammo ca. 1975 and did not have a cannelure. Weight was practically identical to the M1 bullets used in Test C.

1/2" steel plate results:	Identical results with Test C.
Car door B results:	One shot passed through the outer door panel, then both sides of the reinforcing panel. It missed striking the inner panel and created a 3/16" deep crater on the backing plate. One shot penetrated the outer panel, both sides of the reinforcing panel, then left a 1" hole in the inner panel, plus small holes caused by bullet fragments. The backing plate caught only fragments and a lead spray. One shot hit near the top of the door and struck only the outer panel, a thin stiffener panel and the inner panel. This bullet left a 1/8" deep crater in the backing plate.
Comment:	This non-cannelured bullet appeared somewhat stronger than the old .30-06 M1 service bullet made in the 1920's and 1930's but still was far from reliable at this velocity level.

TEST E

Rifle:	Same as used in Tests C and D.
Cases:	Arsenal.
Primers:	W-W 120/8½.
Powder:	Government, probably IMR 4676 or 4895.

Bullet: 165 gr. armor piercing ca. 1943.

Velocity: Not checked.

Steel plate results: For this bullet, a 3/8" steel plate was placed directly behind the 1/2" plate at 200 yards. On impact, the bullet jacket lodged in the 1/2" plate but the bullet core continued on and penetrated the 3/8" plate, then plowed into the hard packed ground with an impressive amount of zip and penetration. In other words, this bullet penetrated a total steel thickness of 7/8" at 200 yards, when launched at about 2750 fps muzzle velocity.

Car door B results: These bullets were very inaccurate and even showed tipping on a 100 yard paper target. One shot went through the outer door panel, then both sides of the reinforcing panel. At this point, the bullet lost stability and its passage through the inner panel was an elongated rip 5/16"x5/8". The bullet hit the 3/8" steel backing plate sideways and left an elongated dent 5/16" wide x 1-1/4" long and 1/8" deep. The deflection from course by this bullet was severe. It appeared to be over 5" in some 18" or less of bullet travel.

One shot was fired to check on the first and produced the same sideways impact, but with even more bullet deflection.

Comment: I had expected great things from this bullet on the car door tests, but two failures out of two were a letdown, to say the least. The course deflection of the bullet after its loss of stability in passing through the car door was astounding. I believe it could miss a small V-6 car engine 3 feet behind a door, even if perfectly directed up to that point. The inaccuracy of the bullets made it difficult to place shots at 200 yards with anything approaching precision. The arsenals making this stuff must have had a quality control department whose only motto was: "Ship it!" I certainly cannot recommend any 165 gr. armor piercing .30-06 government bullets for combat use unless the accuracy is adequate due to that particular batch being a good lot of the stuff. Preferably it should be fired through a custom barrel with a 9" or even quicker twist. Given an accurate lot of bullets and a good custom barrel able to stabilize it, .30 caliber AP could be tremendous. But in other than these conditions, I would not even bother with the ammunition for combat usage.

TEST F

Rifle: 8x57 Mauser military with 23-5/8" barrel.

Load: Winchester 154 gr. FMJ bullet, 53 gr. of ball powder resembling 748. This ammunition had amazing variations in velocity and pressure.

Velocity: From 2790 to 3060 fps.

1/2" steel plate results: At 200 yards, this test resulted in a crater 3/8" deep with a bulge in the back of the plate 1/8" high.

Car door B results: The huge variation in pressure and velocity made it impossible to place shots with any degree of precision. One shot went low on the outer panel and missed everything else within the door. This bullet created a crater 5/8" in diameter and a full 1/4" deep.

One shot landed 4-1/2" down from the top of the door, passed through the outer panel, then a stiffener in the door about .035″ thick and made a crater 1/4″ deep in the backing plate. One shot went through the outer door panel and both sides of the reinforcing panel without showing any sign of expansion. The bullet then hit a corner of the inner panel and made a crater 3/32″ deep in the backing plate. One shot hit the outer panel, then the reinforcing panel where a piece of metal is welded on in an overlap weld, creating a total metal thickness of about .200″. This bullet made a 5/8″x7/8″ tear in the inner panel and created a crater in the steel backing plate 3/4″ in diameter and over 1/16″ deep!

Comment: This is a really tough, stable lead core bullet. Evidently, it was made after vehicles became a factor in small arms planning. The pressure variation of the load was unreal. Even good, stiff handloads, similar in pressure to a working .30-06 load, only expand a Winchester case to about .4690″ diameter. Some of these rounds were the familiar .4690″ diameter and some were .4710″ diameter. What kind of pressure that means is hard to place, but probably over 60,000 lbs. C.U.P. pressure. The recoil with these extra hot loads felt more like a .300 Magnum than an 8x57 Mauser. They should have a muzzle energy in the 3200 ft. lb. range! The primers were the flattest I have ever seen in a factory round round except for some proof loads.

TEST G

Rifle: Sako-Colt .223 with 24″ barrel.

Load: South Korean 55 gr. FMJ, identical to the government .223 round.

Velocity: Not tested.

1/4″ steel plate results: This load proved to be slightly more potent than the .222 loads, as it penetrated this 1/4″ plate consistently at 200 yards.

Car door C results: One shot went through the outer panel, the single reinforcing panel and the inner panel, where it left a .45 caliber hole. The impact on the 1/4″ thick backing plate was 1/16″ deep. One shot made a .40 caliber hole in the reinforcing panel and a .50 caliber hole on the inner panel. The impact of this bullet on the 1/4″ plate was merely a small spray of lead. A third shot also sprayed lead on the backing plate but did leave a crater 1/32″ deep.

Comment: This car door was by far the easiest one to penetrate, yet the action of this load was unreliable. I did a retest with Hornady 55 gr. FMJ Boattail bullets and 22 gr. of 4198 powder and got practically identical results. These bullets are small and unless they remain in one piece, possess very little capability for damage on an engine.

TEST H

Rifle: .308 Norma Magnum Bullgun with 28″ barrel.

Cases: Norma.

Powder: MR 3100 – 72 gr.

Primers: Federal 215.

Bullet: 165 gr. armor piercing, pulled from .30-06 military ammo.

Velocity:	Not checked.
Test results on car door A at 200 yards:	One shot went through the outer panel, then the bottom edge of the reinforcing panel where the two sheets are welded together. The bullet missed the inner panel and the core went through the 1/2" plate where a 3/8" deep crater existed. The hole in the 1/8" steel plate showed evidence of instability. The second shot went through the door but landed sideways on the 1/2" plate, making a crater 3/16" deep.

9

Bullets And Barriers - Automobile Tires

This test provided some surprises. The .222 loads, used in tests 4A and 4B on car doors and steel plates, were tested on a G78-15 tire at both 200 yards and close up. No evidence of expansion could be found. The bullets whipped through both sidewalls as if they were paper. These were varmint-type bullets, too, intended for rapid blow-up on chucks to provide safe shooting in settled areas. Obviously, if a varmint load will not expand on a tire, there was little point in testing heavier calibers with much tougher bullets on a car tire.

The really significant point from this test is that shooting through the upper part of the front tire is the path of least resistance to the engine on a car and the surest way to significantly damage the engine in order to halt the vehicle. Law enforcement officers using rifles firing the .223 cartridge should keep this in mind.

D-P 10 7.62 NATO

Bullets And Barriers - An Overview

Reliable penetration of barriers is a must if a sniper rifle is to attain its highest efficiency. Barriers to a rifle bullet can be of many types. The following is a partial list of them:

A. Trees and logs.
B. Cement blocks.
C. Loose and hard-packed earth.
D. Rocks.
E. Sand or fine gravel in bags.
F. Steel plates of single or multiple thickness.
G. Multiple layers of sheet metal (such as a car door).
H. Bullet-proof glass.
I. Steel shot in bags.
J. Abrasive grain in bags.
K. Ceramic armor backed by fiberglass.
L. Automobile tires.

The first two items investigated were steel plates and car doors as barriers. This series of tests was broken into two parts. The first part evaluated the ordinarily available soft point hunting bullets and the readily available hollow point hunting bullets as well as the readily available hollow point target and varmint bullets. The second part of the testing was done with military style full metal patched bullets, plus armor piercing military bullets and specially made .30 caliber FMJ bullet by Barnes. Some hunting bullets which had been specially modified to give better penetration were also tested.

Over a dozen rifle cartridges were evaluated, ranging from the .222 Remington up to and including the .375 Magnum. In most cases, more than one type and weight of bullet was evaluated in each caliber. Various thicknesses of steel plates were procured and three different car doors. It is not generally known but car doors have a reinforcement panel that may be either one or two layers of metal running lengthwise of the car. This is to give some measure of protection to the occupants of the car in case of a side collision. On the first door tested (designated as door "A"), the combined metal thickness of these two layers of longitudinal reinforcing is just over .140". The inner panel is .034" thick on this particular door. I was unable to get an accurate measurement on the outer panel. The second car door (door "B") also had a total reinforcement thickness of just over .140" and the inner panel was .036" thick. This particular door was from a 1976 Buick Skyhawk, a two door car. The third car door (door "C") was from a 1974 Plymouth Duster. The reinforcing panel was a single layer of metal .070" thick, but it had been formed on a press so that half of the metal was near the outer door panel and some was more than an inch away from the same panel. The inner panel thickness was .040".

When I started this project, I was certainly naive in my expectations as to the capabilities of conventional lead core bullets in penetrating an object such as a car door. I had intended to set up two car doors 6 feet apart and place a steel plate behind the second door to determine, in a rough fashion, the impact energy possessed by the bullet after penetrating both doors. Such an expectation I found is about like a fat man of 50 expecting to run a four minute mile after 6 months of jogging. It soon became glaringly apparent that the lead cored, copper alloy jacketed bullets made in the USA were for the most part far too fragile for the task at hand. I had supposed that a bullet capable of penetrating a mule deer from end to end at 350 yards while expanding nicely in the process would react to a car door like it was a minor hindrance to progress. Such did not prove to be the case!

Likewise, when I started this project I had unbounded faith in the ability of our Ordnance Department to design and manufacture rifle projectiles of optimum penetration reliability on many kinds of barriers. I am now much sadder but wiser. Further, I am convinced that much can be done by the individual experimenter to create combat loads for the sniper rifle that are superior to what is normally available in commercial and military channels. Initial testing was done at 200 yards. The first cartridge tested was an ancient round, the .256 Newton. At one time, this was a factory cartridge but was dropped from production about 1940. The only reason I used this cartridge was because Sweidish 6.5mm bullets can be used in the .256 Newton. These Swedish made bullets have *steel* jackets, not a copper alloy such as in our domestic production. Also, some rifles are available in this country that can use these foreign-made bullets. Examples are: 6.5mm Remington Magnum, the .264 Winchester Magnum and the excellent 6.5x55mm military cartridge. A very good Wildcat can be made by necking down .30-06 or .270 Winchester cases to use the 6.5mm bullets. The performance level of such a round is slightly above the .256 Newton used in the tests.

The data for the tests is as follows:

TEST 1

Barrel length: **24".**

Barrel source: **Made in Germany for the Charles Newton Rifle Corp., Buffalo, NY. Made of rustless steel.**

Cases used: **LC69 .30-06 reformed to .256 Newton. Neck reamed for proper clearance.**

Primer:	W-W 120/8½.
Bullet:	129 gr. Hornady Spire Point (without the interlock feature).
Powder type:	Hodgdon 4831 New Manufacture. This was a hot lot and I suspect it was 1969 government surplus and not the Scottish-made product. 53 gr.
Velocity:	3040 – 3085 fps.
Steel plate results:	Complete penetration of the 3/8" plate at 200 yards with ease. On the 1/2" plate, one shot went through and the other just failed. This shot created a bulge on the back of the plate 1/4" high and partially cracked around the base of the bulge.
Car door results:	One shot went through the outer panel and the inner panel, missing the reinforcing panel completely. This bullet showed almost no evidence of expansion coming through the inner door panel.

One shot went through the outer door panel, then through both thicknesses of reinforcing panel plus the inner door panel. The progress through the door could be correlated with its rate of expansion. The hole in the outer panel was just about bullet diameter. Bullet expansion started on the outer layer of reinforcing panel and an inch away left a 1/2" hole penetrating the inner section of the reinforcing panel. The inner door panel next to the passenger compartment had a hole 3/4" to an irregular 7/8" in size. In examining the path of the bullet, it almost seemed that a cone was formed. Expressed another way, it looked as if once expansion started, it progressed at a certain rate per inch of bullet forward movement.

Comment: This and other car door tests revealed that the presence of the reinforcing panel in a car door is super important in determining what kind of condition a bullet exits the car door. Likewise, the various steel items that raise and lower the windows can have dramatic influence on how a bullet exits the car door.

The second test utilized a Mark X Mauser in .270 Winchester. Test No. 2 data was as follows:

Barrel length: 24".

Case used: DWM.

Primer: W-W 120/8½.

Bullet: 130 gr. Sierra Boattail.

Powder: Hodgdon 205 – 55-1/2 gr.

Velocity: Not checked.

Steel plate results: On the 1/2" steel plate at 200 yards, two shots were fired. Both failed to go through. The crater was over 7/16" deep and the bulge on the back of the plate was 1/4" high. If the velocity had been 100 fps higher the bullets probably would have penetrated.

Car door results: On car door A, the Sierra bullet went through the outer panel, then through both thicknesses of reinforcing panel and made a 1" diameter hole on the inner panel. A 3/8" steel plate had been set up behind the door and the bullet merely made a spray of lead on the plate. No evidence of any type of dent on the steel plate from the bullet impact. It appeared that the .270 caliber, 130 gr. Sierra bullet was somewhat more fragile than the 129 gr. 6.5mm Hornady Spire Point.

Test 2A data used the following:

Cases:	Frontier Cartridge.
Primer:	W-W 120/8½.
Bullet:	150 gr. Sierra Boattail.
Powder:	Hodgdon 4831 N.M., 56 gr.
Velocity:	Not checked.
Steel plate results:	On the 1/2″ steel plate at 200 yards, no shot penetrated. The bulge on the back of the plate was about 3/16″ high or just slightly less than that created by the faster 130 gr. loads.
Car door A results:	These bullets would penetrate all four thicknesses of metal in the car door, but the holes in the inner panel were 7/8″ in size. The impact on the 3/8″ backing plate was a lead spray 3″x4″ except for a small dent in the plate about .050″ deep.
Comment:	This bullet did not penetrate quite as deeply in the steel plate as the 130 gr. bullets, but had a little less tendency for total disintegration on the car door.

Test 3 was conducted using these items:

Rifle:	.270 Winchester custom Mauser.
Barrel:	26″ target weight with a 12″ twist.

Cases: Western Super X.

Primers: W-W 120/8½.

Bullet: 130 gr. Sierra Boattail.

Powder: 4064 –49-1/2 gr.

Velocity: 3135-3160 fps.

Steel plate results: This rifle put its bullets through the 1/2″ steel plate at 200 yards with ease.

Car door results: Not tested.

Comment: This rifle was wonderfully accurate. For some reason, this 25 year old rifle has a fast barrel.

Test 4 had the following history:

Rifle: Winchester Model 70 with 22″ barrel.

Cases: Remington-Peters.

Primers: W-W 116/6½.

Bullet: Hornady 50 SX.

Powder: Hercules Reloader No. 7 – 21 gr.

Velocity: Not checked.

Steel plate results: On 3/8″ plate at 200 yards, these bullets made 3/16″ deep craters and 1/32″ bulges on the back of the plate.

Car door A results: When the Hornady bullet hit the outer and inner door panels, it made a 3/8″ hole on the

Results of the .256 Newton on a 1/2" steel plate at 200 yards. One shot went through and one just failed by a very narrow margin. The 129 gr. Hornady Spire Point delivered a muzzle velocity of 3040-3085 fps.

7mm/06 and .308 Norma Magnum results on 1/2" steel plate at 200 yards. Two out of four 7mm/06 shots went through while all shots from the .308 went through with power to spare.

Left to right: .308 Norma Magnum loaded with 180 gr. Hornady bullet; 180 gr. Hornady .30 caliber Spire Point bullet; fired .223 cartirdge case; 55 gr. Hornady FMJ bullet for the .223. The .308 Norma Magnum will go through a 1/2" steel plate at 200 yards, using the 180 gr. lead core bullets while the .223 will only go through a 1/4" steel plate at that distance. There is something to be said for the higher powered loads.

inner panel and a crater 3/64" deep on the 3/8" plate behind the door. However, when all 4 thicknesses of metal in the door were encountered, the hole in the inner panel was 5/8" across and the bullet only made a small spray of lead on the steel plate.

Comment: At 80 yards. this load made a crater 5/16" deep in the 3/8" plate and created a bulge 1/8" high on the rear of the plate. This shows that steel penetration is sensitive to velocity and that these loads lose it rapidly.

For Test 4A, all components remained the same as for Test 4 except that 50 gr. Remington bullets were used:

Steel plate results: These bullets created a 5/32" deep crater in the face of the plate and a barely discernible bulge on the back of the plate.

Car door results: Not tested.

Test 5 was performed with the following items:

Rifle: Savage Model 340 bolt action in .222 Remington caliber.

Powder: 4198 – 19½ gr.

Bullet: Hornady 50 gr. SX.

Steel plate results: At 200 yards, the bullets made a crater 7/32" deep with a bulge 1/8" high on 1/4" steel plate. On 3/16" steel plate at the same distance, penetration was complete.

Car door results: Not tested.

Comment: This load was accurate and produced less pressure than the 21 gr. load of Reloader No. 7 in Test 4.

Test 6A utilized a .308 Norma Magnum Bullgun with a 28" match grade barrel .

Cases: Norma.

Primers: Federal 215 Magnum.

Bullets: 180 gr. Sierra Soft Point Boattail.

Powder: MR 3100 – 73 gr.

Velocity: 3130-3150 fps.

Steel plate results: At 200 yards this load easily penetrated the 1/2" steel plate. The holes on the exit side of the plate were about 11/16" in diameter.

Car door A results: This test was performed without the steel plate behind the car door. Expansion started on the outside panel of the door and the bullet left a hole over 1/2" in diameter, passing through the second reinforcing panel. The hole in the inside panel, adjacent to the passenger compartment, was 1-1/4" to 1-3/8" across. The same linear progression of expansion first noted with the .256 Newton was noted on this test.

Comment: This bullet was not stopped by any steel plate behind the car door, but if it had been there I am confident that it would have been recovered as a spray of lead and fragmented jacket metal. This load is too hot for steady use and 72 grs. was established as maximum with this bullet weight.

Test 6B employed the same rifle, cases and primers used in Test 6A but with the following variables and results:

Primer:	Federal 215.
Powder:	IMR 4831 – 69 gr.
Bullet:	180 gr. Hornady Spire Point with Interlock Jacket.
Velocity:	3060 fps, approximately.
Steel plate results:	The 1/2" plate was easily penetrated at 200 yards. Holes sizes were slightly smaller than those produced by the 180 gr. Sierra bullets.
Car door A results:	A 3/8" plate was placed about 12" beyond the car door at 200 yards. One shot went through the outside panel and missed the other three thicknesses of metal in the door. When this bullet hit the 3/8" plate, it penetrated completely. When only the inner and outer panels of the door were hit, the bullet also went completely through the 3/8" plate behind the door. When the bullet from this load passed through the four thicknesses of metal in the door plus a section of metal 1/10" thick used to raise and lower the window, the only evidence of its impact on the 3/8" steel back plate was a spray of lead 5" or more in diameter and a very small indentation about .025" deep.
Comment:	If this bullet had been directed at a car motor using this load it would have been very unreliable. Two thicknesses of fender would have made water jacket fracture highly probable. If a section of the frame had been hit, plus one or more thicknesses of fender, the resulting lead spray would have been incapable

of doing much damage except to the distributor cap or ignition wires.

Test 6C featured the same rifle, cases and primers as used in tests 6A and 6B. Variations and results were:

Powder: IMR 4831 – 68 gr.

Bullet: 180 gr. Nosler Partition Jacket.

Velocity: 3040 fps.

Car door A results: This bullet expands faster than the Hornady 180 gr. Spire Point. In passing through the 4 thicknesses of metal in the car door, it left an incredible hole 2-1/2" in diameter in the inner panel. The spray of lead on the 3/8" back plate was about 4"x6". An indentation 1/32" deep x 9/16" in diameter was found on the plate. It was probably caused by the rear portion of the bullet. A check test gave similar results, the only difference was the hole in the inner panel measured 1-1/4" x 1-3/4" and the indentation was 1/16" deep on the 3/8" steel plate.

Comment: I would like to obtain high speed movies of this Nosler bullet coming through the inner panel. I would bet that partition was about 3/4" in diameter and 1/32" thick when it hit the steel plate.

This series of tests using soft point bullets weighing from 50 gr. to 180 gr. at closely similar velocities proved that none were very reliable in shooting through a car door when the maximum amount of potential bullet interference was encountered. In steel penetration, the soft point bullets made holes approximately twice the bullet diameter. Obviously, deeper steel penetration could be achieved if the bullet's energy were expended more forward and less laterally. The other type of bullet commonly available to handloaders is the hollow point target type. Examples include: the 168 gr. .30 caliber bullets made by Hornady, Nosler and Sierra and the 162 gr. 7mm bullets by Hornady.

Test 7A utilized the following:

Rifle:	7mm/06 Mauser with a 26" Varmint weight barrel having a 9" twist.
Cases:	LC 1969, reformed and trimmed.
Primers:	W-W 120/8½.
Powder:	4350 – 53 gr.
Bullets:	162 gr. Hornady Hollow Point Match Boattail.
Steel plate results:	Initial tests on the 1/2" plate at 200 yards involved four shots. Two went through and two just failed by the narrowest of margins. The bulge on the back of the plate was over 1/4" high and partially cracked at the base.
Car door A results:	These bullets blew up after coming through the second thickness of the reinforcing panel and left holes an inch in diameter in the inner panel. Evidence of bullet fragmentation was shown by small tears in the inside door panel around the 1" hole where small pieces of bullet jacket came through an inch or so away from the main portion of the bullet. The steel plate had not been set up behind the door but if it had, there would have been only a spray of lead on it.

Test 7B involved the same rifle, cases and primers as were used in Test 7A, with the following variables and results:

Powder:	MR 3100 – 56 gr.
Bullets:	162 gr. Hornady Hollow Point Match Boattail.
Velocity:	2770 to 2800 fps.
Steel plate results:	On the 1/2″ plate at 200 yards, this load made a crater 7/16″ deep plus a bulge on the back of the plate 1/4″ high.
Car door results:	Not tested.
Comment:	This load had quite decent accuracy at 100 yards, when tested on a target. The pressure is moderate and the powder charge could be increased.

Test 7C utilized the same rifle, cases and primers as Tests 7A and 7B. Variations from the two previous tests included:

Powder:	MR 3100 – 54 gr.
Bullets:	175 gr. Hornady Spire Point Soft Point.
Velocity:	2820 fps average.
Steel plate results:	The craters on the 1/2″ plate were identical to those produced in test 7B, except the bulge on the back of the plate was only 3/16″ high.
Comment:	This load produced more velocity with less powder than load 7B did. Pressure is as much as you would want with a working load. Accuracy is very good.

Test 7D utilized the same rifle, primers and bullets as were used in Test 7A.

Cases:	SL 1943, reformed and trimmed.
Powder:	MR 3100 – 57-1/2 gr.
Velocity:	2900 fps average.
1/2" steel plate results:	This load made a crater 15/32" deep and bulged the back of the plate over 1/4" high.
Car door B results:	This shot went 3 inches below the window sill and only passed through the inner and outer door panels before striking the steel back-up plate. The resulting crater was 1/8" deep but 3/4" in diameter.
Comment:	This hollow point target bullet is very fragile. Its accuracy cannot be questioned but it seems more fragile than the soft point hunting bullets tested to date.
General comment:	Owners of rifles chambered for the .280 Remington or 7mm Express cartridges will be able to assess quite accurately the capabilities of their rifles from the data from these three tests.

More .30-06 rifles are around than any other caliber. Being a one-time military cartridge, it will be chosen by police departments and survival buffs in great numbers. For this reason, considerable effort was expended assessing its capabilities.

Test 8A consisted of the following:

Rifle:	Winchester Model 70, post-1976 design with 22" barrel.
Cases:	Arsenal.

Primers:	W-W 120/8½.
Powder:	MR 3100 – 56 gr.
Bullets:	200 gr. Sierra Spitzer Soft Point Boattail.
Velocity:	2450 fps, average.
Steel plate results:	On the 3/8″ plate at 200 yards, complete penetration was achieved on every shot. On the 1/2″ plate, the craters formed were 3/8″ deep x 5/8″ in diameter. The bulges on the rear of the plate were 1/8″ high.
Car door B results:	These bullets tore holes 5/8″ in diameter coming through the second layer of the reinforcing panel. On the inside panel, they tore holes from 1″x1-3/8″ to 2″ in diameter. The window mechanism components were not struck on these shots. Obviously, the bullets had disintegrated into shrapnel when they struck the inner panel.
Comment:	For some reason this Sierra bullet is quite fragile and mushy. It expands very quickly and probably would be a good deer bullet but is far too fragile for this type of work.

Test 8B utilized the same rifle and cases as Test 8A and involved the following changes and results:

Primers:	Federal 210.
Bullets:	Hornady 180 gr. Spire Point Interlock type.
Powder:	4350 – 57 gr. This initial load proved too hot and was reduced 1 gr. for later tests.
Velocity:	2690-2710 fps.

Steel plate results:	On the 1/2″ plate at 200 yards, this load created a crater over 7/16″ deep plus a bulge on the back of the plate 1/4″ high.
Car door B results:	This load illustrated the difficulty of getting consistent results shooting through the car door. One shot penetrated the outer panel, then both sides of the reinforcing panel plus hitting part of the window raising mechanism, which stopped it. One shot passed through the outer panel, two sides of the reinforcing panel, came through the inner panel in small pieces and sprayed lead over a 3-1/2″ diameter area on the steel backing plate. One shot missed the reinforcing panel but went through the inner and outer panel without any evidence of expansion. When it hit the 1/2″ backing plate, it left a crater 3/8″ deep and a bulge 1/8″ high on the off-side of the 1/2″ plate. One hit the outer panel and the reinforcing panel on a corner. This deflected and broke up the bullet.
Comment:	This load was too high in pressure for arsenal cases but probably would be okay in W-W or Federal .30-06 cases with their greater capacity and hardness. This bullet was very vulnerable to the reinforcing panel and any other metal within the door.

Test 8C also used the same rifle and cases as in Test 8A, but with the following variations and test results:

Primers:	Remington 9½.
Powder:	4064 – 46 gr.
Bullet:	168 gr. Hornady Hollow Point Boattail Match.
Velocity:	2498 fps, average.

1/2″ steel plate results: This load created craters 3/16″ deep and barely bulged the off side of the plate.

Car door B results: One shot hit the outside panel, then penetrated both sides of the reinforcing panel. The inner panel was not hit. This bullet broke into two pieces amd both hit the plate 1″ apart, creating dents 1/32″ deep. One shot hit the outer panel but missed the reinforcing panel. The hole in the inner panel was 5/8″ in diameter. The crater on the steel plate was 5/32″ deep but 11/16″ in diameter.

One shot went through the outer panel, both sides of the reinforcing panel (leaving a .40 caliber hole) and missed the inner panel. This bullet made a crater 3/32″ deep in the backing plate.

One shot passed through the outer panel and struck the bottom of the reinforcing panel where the two sides are spot welded together. This shot missed the inner panel and made a crater 3/32″ deep on the backing plate.

Comment: This load is actually a mild, 200 yard rapid fire match load. Velocities equal to this can easily be achieved by users of the .308 Winchester. Performance of this bullet was very inconsistent on the car door test.

Test 8D again used the same rifle as in the previous three tests. Load and results were:

Ammunition: W-W 180 gr. Power Point Round Nose.

Velocity: 2610-2625 fps.

Steel plate results: Not tested.

Car door B results:	One shot went through the outer panel, then through three thicknesses of metal at the end of the reinforcing panel. The hole in the inner panel was 1/2"x3/4" and the bullet made a double indentation on the backing plate, 1/64" deep. One shot hit the outer panel, both sides of the reinforcing panel, missed the inner panel and hit the back-up plate between two bullet holes. The splash from this bullet bent the burrs formed by the preceding shots.
Comment:	This is a relatively tough bullet. The velocity is well below the 2700 fps quoted in the ballistic tables.

Test 9A consisted of the following:

Rifle:	Custom .30-30 with a 26" barrel having a 10" twist on a bolt action.
Cases:	W-W and R-P.
Primers:	W-W 120/8½.
Powder:	W-W 748 Ball – 32 gr.
Bullets:	Cast Ideal 31141 with gas check.
Velocity:	2150 fps, approximate.
3/8" steel plate results:	Three shots were fired. All created dents 1/32" deep and 3/8" to 7/32" in diameter in the plate.
Car door results:	Two shots passed through the outer door panel, then through both layers of the reinforcing panel, leaving exit holes the size of a dime. Fragments of the bullets only went through the inner panel and sprayed lead on the backing plate.

	One shot went 1-1/2" down from the "elbow edge" of the door. It passed through the outer panel then bent some spot welded thin sheet metal, passed through the inner panel and sprayed lead on the 3/8" backing plate.
Comment:	One shot struck a 3/4" piece of cast iron pipe with a wall thickness of 1/16". The bullet passed completely through both sides of the pipe! Evidently, a bullet that will dent a 3/8" steel plate 1/32" deep possesses considerable wallop.

Test 9B utilized the same rifle as Test 9A, but with the following ammo change and results:

Cases:	W-W.
Primers:	W-W 120/8½.
Powder:	4350 – 37 gr.
Bullet:	Hornady 180 gr. Spire Point.
Velocity:	2030-2050 fps.
3/8" steel plate results:	This load created a 5/32" deep crater 1/2" in diameter, plus a 1/8" bulge on the off side of the plate.
Car door B results:	Two shots were fired. Both went through the outside panel, then both sides of the reinforcing panel and blasted 3/4"x7/8" holes in the inner panel and cratered the 3/8" backing plate 1/32" deep.
Comment:	Amazingly, slowing this bullet way down greatly retarded its rate of expansion as it passed through the car door and the result

was greater dependability of the bullet. The low velocity greatly reduced its penetration in steel compared to 57 gr. of 4350 from the .30-06 or even heavier charges from the .308 Norma Magnum.

Test 10A was as follows:

Rifle:	.33 Wildcat on full length belted .375 or .300 H&H cases with 26" barrel.
Cases:	Winchester Super-Speed .375 H&H.
Primers:	Winchester 120/8½.
Powder:	MR 3100 – 80 gr.
Bullet:	Barnes 250 gr. Spitzer with .032" jackets.
Velocity:	2850 fps, approximate.
1/2" steel plate results:	Complete penetration on every shot. Holes were large. A .375 H&H case could be passed through most holes.
Car door A results:	Bullet went through 5 thicknesses of metal in the door, including the reinforcing panel. The bullet left a 1-1/2"x2" hole in the inner panel and sprayed lead over several square inches on the steel backing plate. One small piece of jacket stuck to the steel backing plate. The bullet started to expand on the first portion of the reinforcing panel and an inch later, left a hole 5/8" to 7/8" in diameter in the second sheet of the reinforcing panel. A portion of the window raising mechanism, 1/10" in thickness, was penetrated and this apparently shattered the bullet, producing the very large hole in the inside panel and the 4"x6" spray of lead on the backing plate.

Comment: Neither Fred N. Barnes nor Randy Brooks, the present maker of Barnes bullets, recommends this thickness of bullet jacket for velocities this high.

Test 10B utilized the same rifle as test 10A, with the following pertinent data:

Cases: W-W.

Primers: Not recorded.

Powder: H4831 (Lot 69) – 78 gr.

Bullet: 245 gr. Barnes experimental with .049" jacket, thinned to .030" at the front.

Velocity: 2825-2850 fps.

1/2" steel plate results: Complete penetration of plate at 200 yards.

Car door A results: One shot went through the outer door panel, both sides of the reinforcing panel, then passed through the inner panel. The impact on the steel backing plate was 3/16" deep. Another shot passed through the outer panel, both sides of the reinforcing panel, then struck a 1/16" thick plate that carried the handle for raising and lowering the window. The hole in the inner panel was 1-1/4" across. This bullet left only a lead spray on the steel back-up plate. One shot passed through the outer panel but was too low to hit the reinforcing panel and the inner panel. This hit almost went through the 1/2" backing plate. The bulge on the back of the plate was over 1/4" high.

Comment:	The shot that made the 3/16" deep impression in the backing plate would have penetrated the water jacket on a car engine, provided a reinforcing rib or locating pad was not struck.

Test 11 included the following data:

Rifle:	.33 Wildcat on .30-06 case with 27" barrel.
Cases:	W-W.
Primers:	Either W-W 120/8½ or Federal 210.
Powder:	4350 – 61 gr.
Bullet:	Barnes 250 gr. with .032" jacket.
Velocity:	2610-2625 fps.
1/2" steel plate results:	The craters formed at 200 yards were 7/16" deep and the bulge on the back exceeded 1/4" in height.
Car door A results:	One shot hit the outer panel, both sides of the reinforcing panel, then left a 5/8" hole in the inner panel and a 3/16" deep crater in the 3/8" backing plate and a bulge on the back of it 1/8" to 3/16" high. One shot struck the window guide after exiting the second sheet of the reinforcing panel. This bullet made a large tear in the inner panel and sprayed lead on the plate but made no dents.
Comment:	This bullet is just about the right structure for this high velocity but is inadequate for speeds 10% higher. This test and tests 10A and 10B proved that even a 250 gr. bullet can be torn into fragments by the metal within a car door. Few soft point bullets are tougher than these

250 gr. Barnes bullets, so some other design is needed to be reliable in car door penetration.

Test 12 consisted of the following:

Rifle:	Remington 700 BDL with 26″ barrel in .375 H&H Magnum.
Cases:	W-W 250.
Primers:	CCI 250.
Powder:	4064 – 70-1/2 gr.
Bullet:	Hornady 270 gr. Spire Point.
Velocity:	Not checked.
1/2″ steel plate results:	Complete penetration at 200 yards, leaving very large holes.
Car door B results:	Not tested.
Comment:	This rifle had two old Redfield 2-3/4X scopes in G&H mounts with the rifle. Both scopes were unable to hold zero or be re-zeroed. Either the recoil of the rifle wrecked the scopes or the several thousand miles the rifle had traveled on hunting trips in recent months did it. As a result, shots could not be placed with enough precision for the car door test.

Test 13 data and results were:

Rifle:	Winchester Model 88 in .308 Winchester with 22″ barrel.
Cases:	LC National Match.

Primers:	Remington 9½.
Powder:	4064 – 41 gr.
Bullet:	Hornady 180 gr. Spire Point.
Velocity:	2448 Instrumental velocity.
3/8" steel plate results:	At 200 yards, this bullet made a crater over 5/16" deep and made a bulge on the back 1/4" high.
Car door results:	Not tested.

Test 14 consisted of the following:

Rifle:	Remington 40X in .22-250 with 27-1/4" barrel.
Cases:	Norma.
Primers:	Remignton 9½.
Powder:	4350 – 39.4 gr. (a very hot load).
Bullet:	Hornady 60 gr. Hollow Point.
Velocity:	3585 fps.
3/8" steel plate results:	At 200 yards, this load penetrated the plate with ease.
Car door results:	Not tested.
Comment:	Tests 13 and 14 establish beyond a doubt the fact that velocity is the chief factor where single plate penetration is concerned. 38 gr. of

4350 delivers 3400 fps muzzle velocity in the .22-250 listed above. Accuracy of the rifle used in Test 14 was extremely good.

D–P 11 7.62 NATO

Bullets And Barriers - Concrete Blocks

This series of tests, done at 200 yards, was undertaken with a number of rifles ranging in muzzle energy from 1300 to over 4500 ft. lbs. Bullet weights varied from 55 to 250 grs. The concrete used in the blocks was quite coarse and contained stones nearly as large as golf balls. In testing, the blocks were stood on end with the flat sides perpendicular to the impact of the bullet. Details of the tests are as follows:

TEST Ia

Rifle:	Winchester Model 70, post-1976 type, in .30-06.
Cases:	Arsenal.
Primers:	Federal 215.
Powder:	MR 3100 – 57-1/2 gr.
Bullet:	Sierra 200 gr. Soft Point Boattail.
Velocity:	2500 fps, approximate.

Results: The bullet hit left of center, just above the center rib reinforcing. A hole the size of a tennis ball and extending to the edge of the block was blasted into the near side of the block. Amazingly enough, the bullet shattered and only bits of jacket, lead core and ground up concrete struck the rear wall of the block. There was no evidence of cracks in this rear wall. The second shot hit just below the center reinforcing panel in the block. This shot cracked the block but again, the bullet did not penetrate the rear wall of the block.

This bullet is obviously quite fragile in spite of its considerable weight. This characteristic of the bullet also showed itself during the car door test.

TEST Ib

Rifle: Same as Test 1a.

Cases: Same.

Primers: Federal 210.

Powder: 4350 – 55 gr.

Bullet: Barnes 190 gr. FMJ experimental bullet.

Velocity: Not tested.

Results: Slight damage on left center of block. This bullet has an open base and this appears to detract from the structural integrity on a target of this type.

TEST Ic

Rifle: Same as in Test Ia.

Cases: Same.

Primers: Federal 210.

Powder: 4350 – 56 gr.

Bullets: 180 gr. Hornady Spire Point Interlock.

Results: This bullet broke the concrete block into 5 or 6 pieces, which fell to the ground very close to where the block stood. There was no scattering of the pieces as was done with the more powerful rifle.

TEST II

Rifle: Colt-Sako .223 Remington with 24" barrel.

Ammunition: South Korean duplicate of GI 55 gr. FMJ load.

Velocity: Not tested. Probably about 3300 fps.

Results: This bullet would only penetrate one side of the block and then would spray the rear wall of the block with a mixture of fine lead spray, ground up jacket metal and concrete dust. The holes made were quite small. Beyond 150 yards, it would be very unlikely that any semi-automatic .223 could hit the same hole to get complete wall penetration.

TEST III

Rifle: .308 Norma Magnum Bullgun with 28" barrel.

Cases: Norma.

Primers: Federal 210.

Powder: MR 3100 – 72 gr.

Bullet: 180 gr. Hornady Spire Point Interlock.

Velocity: 3100 fps, approximate.

Results: This load shattered the concrete block into a minimum of 8 pieces. One piece was found over 4 feet from its original location. Obviously, this relatively powerful, high velocity rifle was enormously more destructive than the .223 Remington on concrete blocks.

TEST V

Rifle: 7mm/06 Mauser with 26" Varmint weight barrel with a 9" twist.

Cases: Arsenal, reformed and trimmed.

Primers: Federal 215.

Powder: MR 3100 – 54 gr.

Velocity: 2820-2830 fps.

Results: This bullet hit the top half of the block above the center rib and shattered the top portion of the block, but did not shatter the bottom portion as the .308 Norma Magnum had done. Actually, it appeared that this load created less severe destruction than the Hornady 180 gr. did from the .30-06. I suspect that the frontal area of the bullet as well as its impact kinetic energy determine its destructiveness on concrete blocks. Later tests confirmed this.

TEST VI

Rifle: .33 Wildcat on full length Belted H&H cases.

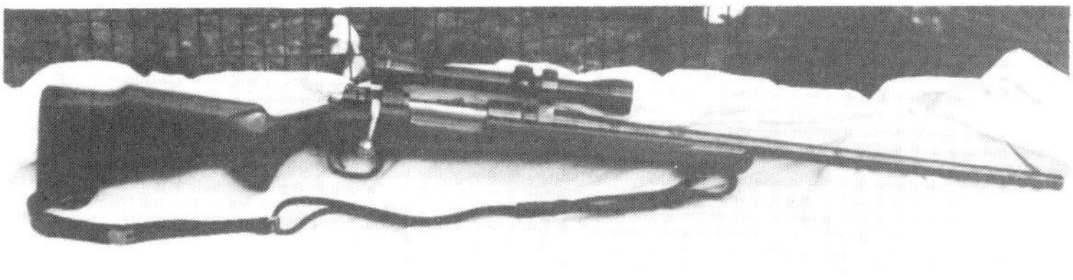

By far the most destructive results on concrete blocks were obtained with a .33 caliber Wildcat cartridge based on the full length belted .300 H&H. The rifle for this load is shown above.

Cases: .300 H&H Western Match.

Primers: Federal 210.

Powder: H 4831 (lot 69) – 79 gr.

Bullet: 250 gr. Barnes Spitzers with .032" jackets.

Velocity: 2875 fps.

Results: This load went through the concrete block and simply exploded it over an area some 10 feet in diameter. 5 or 6 people witnessed this shot from directly behind the shooter and there were gasps of amazement at the results. One observer who has a large collection of game trophies from Africa, Canada and Alaska, commented that if someone were behind a concrete block wall hit with this load, if the bullet didn't get him, the chunks of concrete would!

TEST VII

Rifle: .33 Wildcat on .30–06 case with 27" barrel.

Cases: W-W.

Primers:	Federal 210.
Powder:	4350 – 60.7 gr.
Bullet:	Barnes 250 gr. Spitzers with .032" jackets.
Results:	This load proved just about midway between the .308 Norma Magnum and the large case .33 Wildcat (Test VI) in destructive potential on concrete blocks. It shattered a block completely with every shot but did not blow the fragments over quite as large an area. However, it was very noticeably more destructive than the .308 Norma Magnum.
General comment:	I had intended to try the .375 H&H Magnum on the concrete blocks but when both scopes were found to be out of commission, this idea had to be abandoned. I believe that the 270 gr. Hornady at 2750 fps or so would be at least the equal of and very likely superior to the large case .33 Wildcat used in Test VI in concrete block destruction.

At close range, a .223 semi-automatic rifle would chew through a concrete block wall in short order. However, at longer ranges like 400 yards or more, a .223 would be a long time getting through a concrete block wall, particularly if a heavy wind was drifting the bullets badly. Personally, I believe a SWAT team might find a powerful .30 caliber Magnum or even a .375 H&H a useful addition to their available weaponry. The individual citizen, concerned with armament for a survival retreat, should also give serious consideration to adding such a rifle to his centerfire rifle list.

One particularly noticeable result of these tests was that if a small, fragile bullet hit a large stone closely held by the concrete matrix, the bullet would shatter with very little destruction to the block.

I have shot at rocks the size of a football over a 30 year period with various rifles at ranges of from 150 to probably 350 yards. I have noticed that when a heavy 250 gr. (or larger) bullet hits, they are likely to shatter into fragments. When a 125 gr. Sierra from a .30-06 hits, all you get is a small chip off the rock and a spray of rock dust. The small bullets from varmint rifles do even less damage. The concrete block test results come as no surprise to me.

D–P

12

7.62 NATO

Tests With Modified Soft Point Bullets

Full metal jacket bullets are not always available in many calibers while really excellent quality soft points are available from several makers. After a discussion of this situation, the publishers of this book threw down a challenge. Could I modify a soft point bullet so that it would be more effective on car engines and car doors? In other words, would this modified soft point bullet be equal or superior to full metal jacket bullets in the same caliber?

The 180 gr., .30 caliber Hornady Spire Point with the Interlock feature was selected as the bullet to work with. The alterations consisted of the following:

A. Cut off the soft point entirely.
B. Contour the jacket at the front so that from a side view it resembled a barn roof.
C. Drill a cavity in the bullet nose 3/16" deep.
D. Make a miniature steel point shaped like a mushroom with a droopy outside diameter to match the contour machined on the front end of the bullet. The stem of the steel tip is a light press fit in the drilled hole in the front of the bullet.
E. Harden the steel tip to 45-50 Rockwell C.

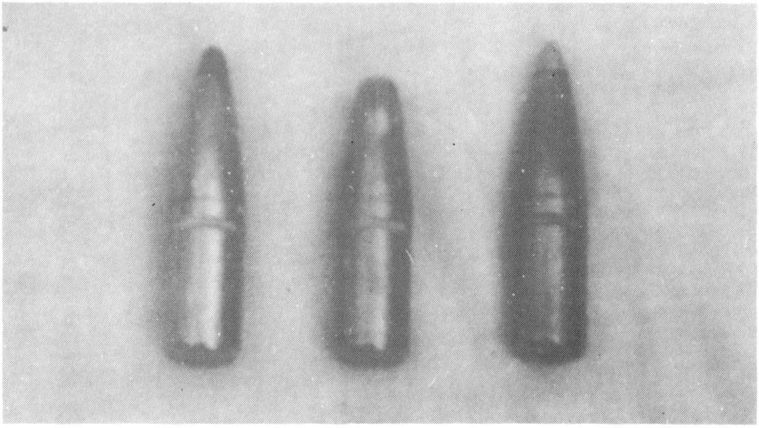

Modification steps taken to turn the 180 gr., .308 Hornady Spire Point Interlock bullet into a potent sniper round included cutting off the soft point entirely, machining the jacket to a "barn roof" contour, drilling and inserting a steel tip.

F. Press the steel tip in the drilled cavity in the bullet nose.

The angle on the front end of the bullet jacket and the bottom side of the steel tip forces the jacket to squeeze together around the stem of the steel tip and resist expanding when the bullet strikes a barrier.

The press fit cannot be excessive or the bullet jacket will fracture in the stress lines induced by the manufacturing process at Hornady's. Probably a bullet with a heavier jacket at the nose would be easier to work with. An example of this type of bullet is the Barnes line of copper tubing jacketed bullets.

In any event, this style of bullet showed improved effectiveness when compared to the unaltered bullet. Test results were as follows:

TEST 1a

Rifle: **.308 Norma Magnum Bullgun with 26" barrel.**

Cases:	Norma.
Primers:	Federal 210.
Powder:	MR 3100 – 72 gr.
Bullet:	Steel tipped 180 gr. Hornady Spire Point.

Match Hollow Point Boattail target bullets proved in tests to be just as fragile as soft point hunting bullets, since the jackets are quite thin and they come apart readily. From left to right are: 190 gr. .30 caliber, 168 gr. .30 caliber and 162 gr. 7mm, all from Hornady. The difference between these bullets' penetration performances and that of the modified 180 gr. Spire Point detailed in this chapter was dramatic to say the least.

At left is the .30 caliber 180 gr. Hornady Spire Point bullet with the soft point cut off, the jacket contoured on the outside and the bullet drilled to accept the steel tip. This bullet split when the steel tip was forced in due to too much interference. At right is a completed bullet of the same type, with the steel point in place.

Velocity:	3100 fps.
Test results at 200 yards:	On the engine block, this bullet penetrated a 1/16" steel plate in front of the engine, then zipped through the water jacket and knocked a hole in the side of the cylinder 1–1/2" in diameter, where the cylinder wall is over 1/4" thick.
Comment:	This amount of destruction was equal to that done by the .375 H&H Magnum, 300 gr. Full Metal Patch bullet at 25 yards. This damage

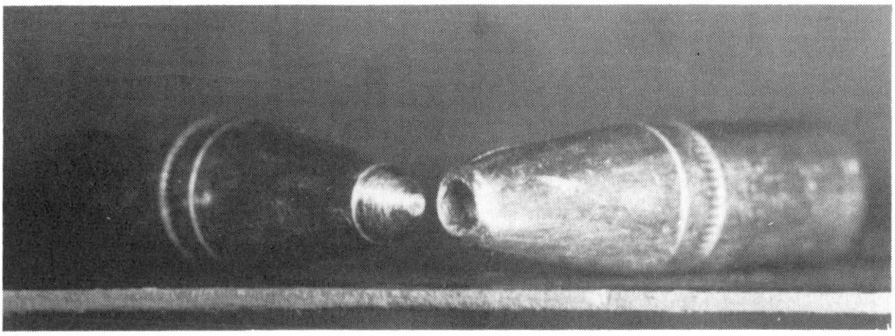

The modified Hornady 180 gr. Spire Point can be seen in its final form to the left, withe the steel tip in place. Modification steps are listed in order in the text. The final bullet is similar to the Remington Bronze Point in design but the special matching taper on the steel point and the bullet nose retards expansion and gives much better results on an engine block.

would have stopped an operating engine very abruptly.

TEST 1b

Rifle:	Same as for Test 1a.
Load:	Same as for Test 1a.
Test results at 200 yards:	On car door B, one shot hit the outer panel, missed the reinforcing panel and the inner panel, then went through the 3/8" plate. One shot went through the outer door panel, both sides of the reinforcing panel, passed through the inner panel and sprayed lead on the backing plate.
Comment:	The steel tip provides some delay in expansion and disintegration, but the reinforcing panel in this car door is made of quite strong material. A heavier jacket at the nose would have made a better showing.

TEST 2a

Rifle:	.30-06 Winchester Model 70, post-1976 type.
Cases:	Arsenal.
Primers:	Federal 210.
Powder:	4350 – 56 gr.
Bullet:	Steel tipped 180 gr. Hornady Spire Point.
Test results at 200 yards:	On car door B, this bullet went through the outer panel, missed the reinforcing panel, passed through the inner panel and cratered the 3/8" backing plate 11/32" deep, plus a 1/4" high bulge on the back of the plate.

TEST 2b

Rifle:	Same as Test 2a.
Load:	Same as Test 2a.
Test results at 200 yards:	On the 1/2" plate, this bullet penetrated the plate. The most interesting thing about this shot was that the axis of the hole in the plate was about 30° or more from the axis of the bullet trajectory. Either the point skidded when it hit the plate or the bullet was yawing in flight.
Comment:	A comparison shot fired just before this one, but with an unaltered bullet, went slightly over 3/8" deep in the plate.

13

Experimental Testing With Alternate Armor Piercing Bullet Design

The 165 gr., .30-06 armor piercing bullet has a steel core weighing approximately 81 gr. This core has a diameter of .244" to .245" and a length of 1-1/16". The base of this core has a slight boattail and the point has about a .04 caliber ogive. In size and shape, they resemble a 6mm 100 gr. bullet with a rather blunt Spitzer point and a limited boattail. The hardness of those checked ranged from 48-55 Rockwell C.

A number of concepts were evaluated to create an armor piercing bullet that the person with a home workshop could make with rather limited equipment. The following information is a chronology of the thinking and experiments performed towards achieveing this goal.

My initial reaction to the design of the 165 gr. armor piercing bullets was that it made very little sense to propel a 165 gr. bullet at 2800 fps when the effective payload weighed only 81 gr. The first idea tried was to make a tool steel bullet with grooves on the rear portion of the bullet where driving bands would be shrunk in place. This would be similar to artillery projectiles. The core of an artillery projectile is very hard and has several copper driving bands to permit the rifling grooves in the barrel to grip the soft copper bands and impart spin. The copper bands also seal the bore against the propelling gases. In short, the hard steel of the artillery projectile does not contact the barrel of the artillery piece.

The diameter of these copper driving bands is closely held. In 1951, I worked for a short time at an Army Ordnance Depot, reworking 90mm artillery shells. One portion of the rework consisted of grinding the driving band to a certain size on a centerless grinder set up for infeed operation. The tolerances were fairly close as I remember and varied about 1/1000" from high to low permissable dimension.

I had figured that the grooves in a high production set-up for my design of A.P. bullet would be ground in like the grooves in an automobile exhaust or intake valve stem are ground in. This valve grooving operation is generally done with 80-90 grit premium aluminum oxide, vitrified grinding wheels of very hard grade, using grinding oil as a coolant. The actual wheel contact time to grind these grooves in the valve is 2-3 seconds.

If such technology was applied to making A.P. bullets, a wheel 6" wide could grind 4 bullets at one time and the cycle time per infeed would be about 6-8 seconds. This would mean a production rate of 18-24 bullets grooved per minute.

The question arose: What could be done along the line of making bullets of this design, using a lathe and drill press? We attempted to make up such driving band bullets from 5/16" rods of 01 tool steel using unhardened material and a lathe to cut the grooves and get the proper diameter. Then the parts were heated in a furnace and given an oil quelnch. The maximum diameter of these bullets was .295". To our surprise the heat treatment created a severe banana shaped distortion of these lathe-turned projectiles. This distortion was of such magnitude that the steel projectiles of .295" diameter would not pass through a normal .30 caliber barrel having a .300" diameter!

The second problem arose when we tried to swage driving bands on the bullets without using a Torrington Rotary Swager. In spite of our best efforts, we simply could not do it. We even tried using a collet and similar systems. Even after annealing the copper to a dead soft condition, we could not make it go. We finally gave up and went to a core, shaped somewhat like a mushroom, with a cup-shaped brass jacket that was a press fit over the rear shank of the steel core. These bullets were made with a hard steel core and front portion and had a bore-riding section of soft brass. The hard steel core never contacted the barrel. The brass bore-riding section was grooved to reduce the force required for the rifling to engrave the soft brass and helped reduce pressure . . . . .

Since there was no give at all to the hard steel core under the driving bands, I abandoned the 1/8" wide driving band design and went to narrower grooves and bands 1/16" wide. I forgot to tell the maker of the bullets to make the rear band at least 1/16" wide. The shop man made them up with the rear band .015" or so wide. The front portion of the bullet was quite pointed and the overall bullet length was 1.40".

Details of the test performed are as follows:

TEST 1

Rifle: .308 Norma Magnum Bullgun with 28" barrel.

Case: W-W .300 Winchester Magnum, reformed.

Primers: Federal 215.

Powder: MR 3100 – 72 gr.

Bullet: Special 157-158 gr., 1.40" long, with sharp conical point 7/8" long. Tool steel point and core. The brass driving section had a .002" press fit on the rear shank of the bullet.

Test material: 1/2" steel plate at 200 yards.

Velocity: Not checked.

Results: Penetration of the plate was accomplished with ease and the hole was very slightly over .30

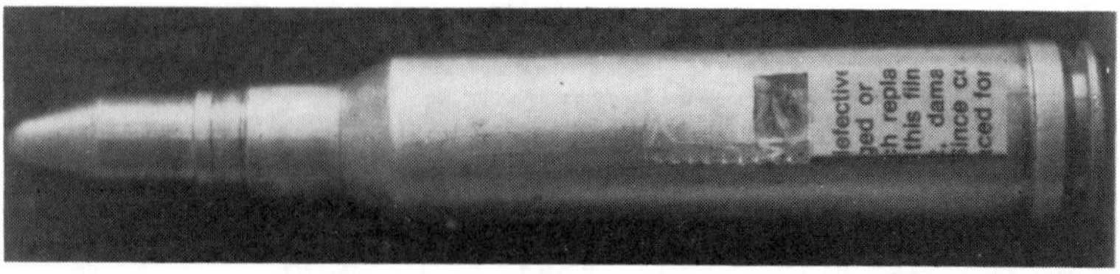

.308 Norma Magnum with special A.P. bullet of narrow land design.

caliber. Yawing of the bullets was severe. Entrance holes indicated the axis of the bullet was 10% off from the primary direction of flight. Accuracy was very poor, far too poor to place shots with any degree of accuracy.

Comment: These bullet cores were ground from solid rod after heat treatment to eliminate distortion. It appeared that the yawing and poor accuracy were the result of too long a bullet for stability. The bullet design was modified for the next test.

TEST 2

Rifle: Same as Test 1.

Cases: Norma.

Primers: Federal 215.

Powder: H4831 (lot 69) – 74 gr.

Bullet: Special 152-157 gr., 1-5/16" long with a cone point and rounded end on the point.

Velocity: Not checked.

Test material: 1/2" plate, backed up by 3/8" plate.

Results: The first shot hit 12" to the right and 3" high at 200 yards. The bullet went right through the 1/2" plate but missed the 3/8" plate behind it, hitting a 1/4" plate instead.

The second shot hit the point of aim and passed through the 1/2" plate, leaving a .32-.35 caliber hole. This bullet struck the 3/8" plate, making a dent 1/8" deep and slightly bulging the back.

The third shot hit low and some 10" to the left of the point of aim, striking the dirt in front of the assembly of steel plates.

Comment: The hardened steel cores of both these bullet designs weighed at least 50% more than the core of the 165 gr. military A.P., yet penetration was less. Accuracy was so poor that even the sad military A.P. bullets looked good by comparison. Either bullet was tough enough that it would have severely damaged any automobile engine unless a heavy boss was struck. The accuracy was so poor that these bullets would be worthless for sniping, however.

The poor accuracy of this design of bullet had to be dealt with. After considerable thought, I decided that the very thin rear driving band was being distorted or even burned badly by the powder gases and this was causing the inaccurate results.

To solve this problem I designed a bullet with driving bands similar to the brass wood-penetrating bullet and with an unhardened steel core. This bullet is illustrated in Figure 1.

To obtain the utmost precision, I had them made by the most precise shop I knew of in my area. The owner of this shop did prototype work for MIT some 40 years ago and later became a mechanical engineer. His work is of the highest order. He made a small quantity for testing at a cost exceeding $15.00 per bullet.

I shot 6 of these expensive bullets at 200 yards from the .308 Norma Magnum and got very interesting results as shown in Test 3:

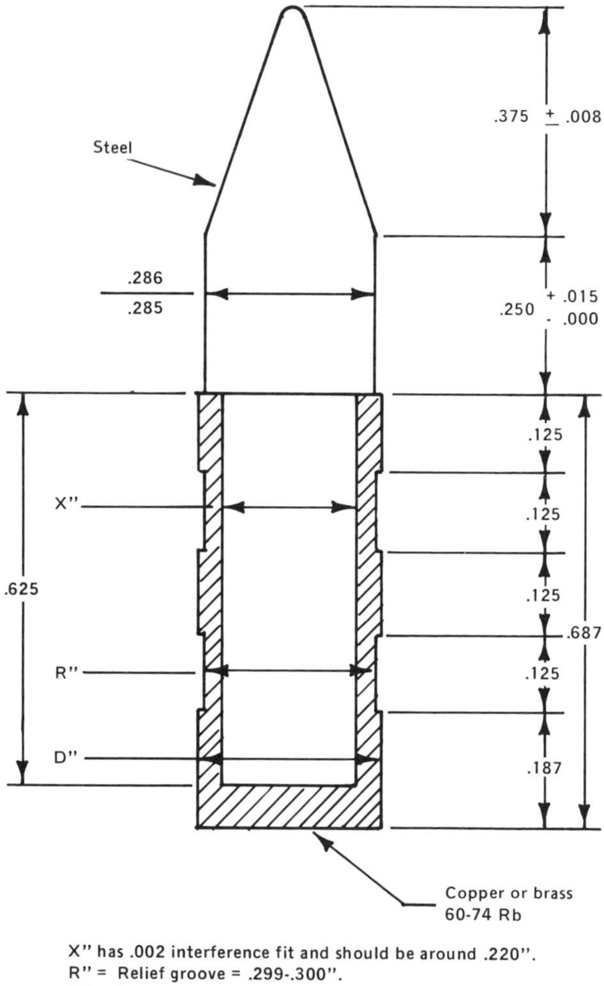

X" has .002 interference fit and should be around .220".
R" = Relief groove = .299-.300".
D" = Driving band @ .309-.3088".

FIGURE 1

Dynamic balance is a must since the velocity of this bullet is 3200 fps and the rifle barrel twist is 1 turn in 10". Uniformity of wall thickness on copper should be around .0002" for accuracy. Steel can be 1020-1045. The .125" dimension on the relief grooves and bore riding bands is nominal. The important thing is to make them uniform from bullet to bullet.

WARNING!

On bullets such as the one shown on the opposite page, IT IS IMPERATIVE THAT THE STEEL FORWARD SECTION BE SMALLER IN DIAMETER THAN THE BORE OF THE RIFLE. If the steel is allowed to come in contact with the bore, severe excess pressure may develop, resulting in the destruction of the rifle and possibly causing serious injury or even death for the shooter.

TEST 3

Rifle: .308 Norma Magnum Bullgun with 26" barrel.

Cases: Norma.

Primers: Federal 215.

Powder: H4831 (lot 69) – 74 gr.

Bullet: Special make, mild steel core, of .285" maximum diameter, brass base .687" long with 3 wide driving bands of .308" diameter. Weight was 154 gr.

Velocity: Not checked.

Test material: 1/16" plate with a target fastened to it for an aiming point. One foot behind that was the V-8 auto engine block used in earlier tests. Behind the engine block was a 1/2" steel plate.

Test results: Accuracy was extremely good. 5 shots grouped under 2" at 200 yards. One "flyer" made the total group 3-1/2" at 200 yards.

The first two shots fired proved very instructive. One shot went through the 1/16" plate, the engine water jacket at the 1/10" thick point and the cylinder wall on both the near and far sides. The second shot hit the smooth top of the block where the cylinder head bolts on and left a skid mark 5/8" long. This deflected the bullet over 45° from its original course and it struck the dirt to the side of the target. When recovered from the dirt, the bullet was badly shattered by either the impact with the engine block or with small rocks on the ground.

The next two shots landed 1-1/16″ apart and less than an inch from the shot that glanced off the cylinder head area. These bullets entered an open cylinder and passed through the far cylinder wall and the metal beyond where the valve tappets ride up and down on the camshaft. The destruction was quite impressive. The fifth shot was fired with the aiming point shifted up some 3″. This bullet went through the 1/16″ plate and struck a very heavy mounting boss on the water jacket with a glancing impact. This bullet chipped away a small corner of the boss and deflected without even cracking the water jacket. It hit 13/16″ high and 3/4″ to the right of the point of aim at 200 yards.

For the last shot, the engine block was moved away so that the bullet would impact on the 1/2″ steel plate. This bullet went through the 1/16″ plate and made a crater 1/2″x5/8″ and about 1/4″ deep on the 1/2″ plate. The cavity was thoroughly plated with brass from the driving portion of the bullet.

The final shot was fired at a live Birch tree some 14″ in diameter and failed to penetrate it.

Comment: The extremely precise manufacture made these bullets very accurate when driving bands were used that were not deformed by the propellant gases. The soft core made the bullet worthless against heavy plate and the sharp conical point made it glance when sharply angled smooth surfaces were struck. The lack of a flat portion on the point apparently hurt its wood penetrating ability. If this type of jacket had been used with the core of the

Left: 7mm bullet with Stellite forward section and brass bore riding section containing 3 wide bore riding lands. This bullet proved very accurate. Center: Hard steel forward section bullet with narrow band brass bore riding section. This bullet gave good penetration but poor accuracy. Right: .30 caliber soft steel core bullet with brass lands. This bullet gave very good accuracy but poor penetration.

type used in Test 2, an accurate and very deadly bullet would have resulted.

TEST 4

Rifle: 7mm/06 with 26" Varmint weight barrel.

Cases: Arsenal.

Primers: Federal 215.

Powder:	H4831 (lot 69) – 56 gr.
Bullet:	Special make, Stellite core of .255″ maximum diameter, brass rear section with 3 wide bands. Overall length of the bullet was 1–3/8″. The tapered point was 7/16″ long with a flat point 1/16″ in diameter. Bullet weight was 133 gr.
Velocity:	Not checked.
Test material:	1/2″ steel plate backed up by a 1/4″ steel plate.
Test results:	Two shots were fired. One failed to penetrate the 1/2″ plate by the narrowest of margins and became lodged in it. The other shot penetrated the 1/2″ steel plate and lodged in the 1/4″ plate. Interestingly, the second shot's axis of penetration was some 20° from its line of flight. Neither bullet core showed much evidence of damage from impact with the steel plates.
Comment:	In external length and shape, these bullets rather closely resembled a 175 gr. Hornady 7mm Spire Point bullet. The weight with the Stellite core turned out to be disappointingly low. Machining the Stellite material was very difficult. Its resistance to impact damage was impressive. The cost of the Stellite rod was about $1.00 per inch.

14

Tests With Special Homogenous Turned Bullets

Tests were conducted using bullets turned from 1/2" hard brass. The bullets were of two types. One had a conical point 3/8" long and a flat end about 1/10" in diameter. The other had a steel tip of conical shape attached to the flat nose of the bullet. This bullet resembled the Remington Bronze Point in appearance. The flat ended bullets were slightly over 1-1/4" long and those with the steel tip were 1-7/16" long. Weights were 160-162 gr. for the flat end bullets and about 165 gr. for those with the steel tip. Just behind the conical point was a .300" diameter cylindrical section 1-1/4" long. Behind that was a driving band 1/8" long of .308 diameter. After that was a 1/8", .300 diameter relief groove followed by another 1/8" .308 driving band, then another 1/8" long relief groove. The base of the bullet had a driving band 5/32" long and a cavity in the base 1/32" deep and 1/4" in diameter.

The purpose of this design was to give a .30 caliber contact on the top of the lands but only have some .400" of bullet bearing on the bottom of the grooves, plus a hollow base to promote gas sealing. This bullet design minimizes pressure in the rifle and gives maximum course-holding ability in shooting through trees. Certain forms of bronze are slightly denser than brass. I had a steel tipped version made in man-

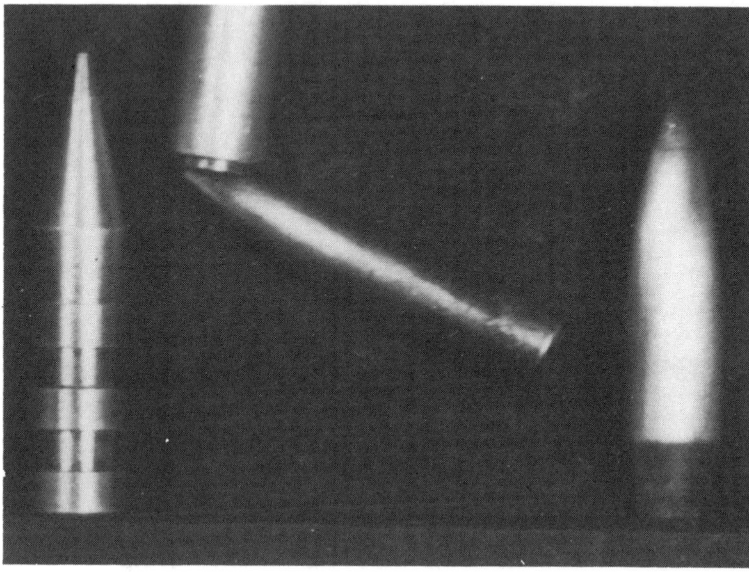

Left to right: .30 caliber lathe turned bullet given a tool steel point; 6.5mm (.264) 139 gr. Norma FMJ Boattail with a steel jacket (suspended from magnet); .30 caliber 180 gr. Sierra Spitzer Soft Point Boattail. The steel tipped lathe turned bullet was a bit long for good stability from a barrel with a 10" twist.

ganese bronze that weighed 178 gr. The hardness of the brass or bronze should not be excessive. I have not worked out just how hard it can be without creating dangerous pressures, but 65 Rockwell B has been okay on 165 gr. brass. The manganese bronze bullet was 77 Rb in hardness and gave pressures about like a lead core bullet of 190 gr. Relief grooves are mandatory to hold down pressures, in my opinion.

The steel tipped version at 1-7/16" in length was too long to be well stabilized, even when fired from the .308 Norma Magnum. This style of bullet possesses unusual ability to shoot through trees and does fairly well going through car doors and damaging engine blocks.

In steel penetration, this bullet is not as good as lead core bullets of similar weight driven by the same charges. The craters they create are more shallow than those of ordinary lead core bullets and are at least as wide, if not wider. Possibly the density of the lead core is a factor in steel penetration. Both brass and bronze are considerably less dense than lead. Their primary advantage is their ability to maintain a course in car body sheet metal and wood. They do not break up on sheet metal as most lead core bullets do. Their advantage over most military bullets in penetrating trees has to be seen to be believed.

Test results were as follows:

TEST 1

Rifle:	.30-06 Winchester Model 70 of post-1976 design with 22" barrel.
Powder:	4350 – 56 gr.
Primers:	Federal 210.
Cases:	Arsenal.
Bullet:	Flat ended version of solid brass design.
Velocity:	Not checked.
Test results at 30 yards:	On a section of a living Cherry tree, this load went straight on through like a hot knife through butter and left a clean, round hole in a 1/16" steel plate behind the tree. Tests with a military FMJ Boattail went through the tree but were traveling at an angle on exit and possessed very little remaining energy.

TEST 2a

Rifle:	.308 Norma Magnum with 28" barrel.
Powder:	MR 3100 – 72 gr.

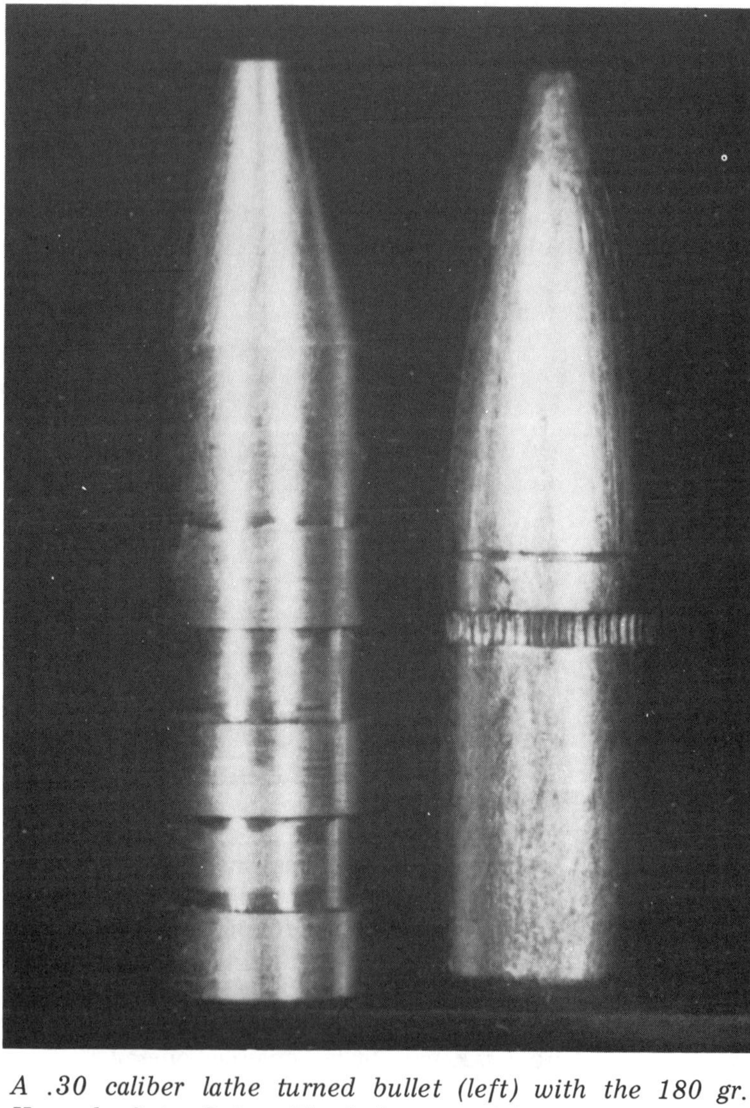

A .30 caliber lathe turned bullet (left) with the 180 gr. Hornady Spire Point. The lathe turned bullets weigh about 164 gr. in brass and 175 gr. in manganese bronze. Penetration of the lathe turned bullet is extremely good in wood and also proved effective in the engine block test. Steel plate penetration was rather poor, however.

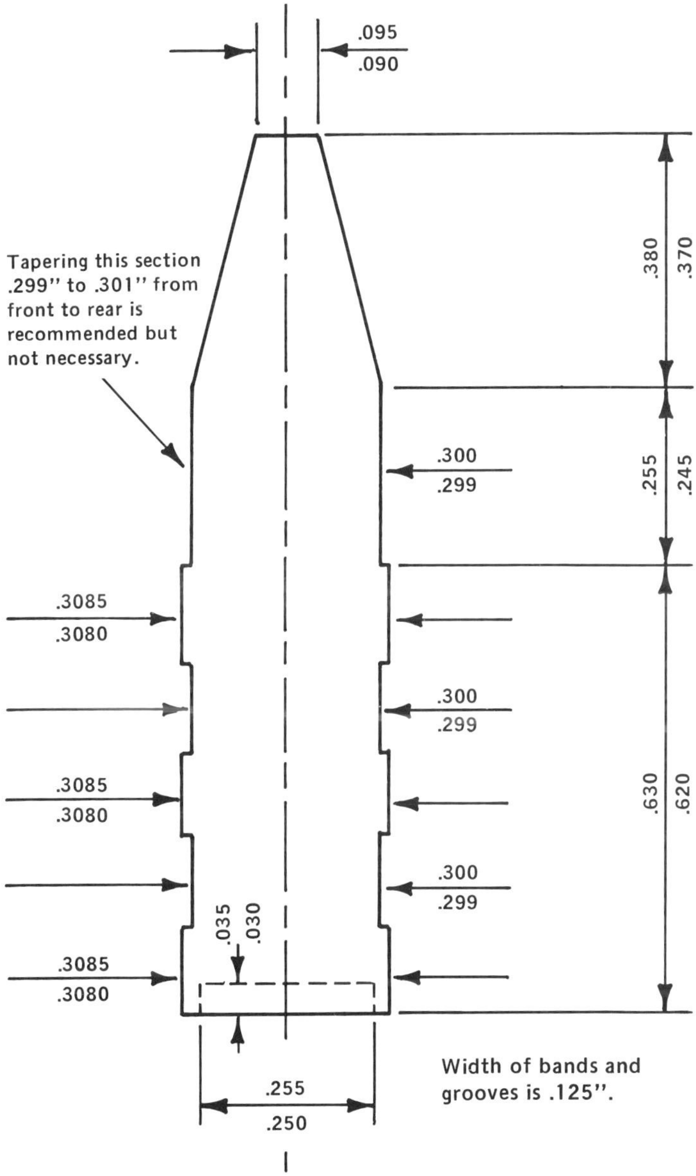

FIGURE 2
Manganese bronze or brass lathe turned bullet.

Base view of the lathe turned bullet (left) clearly shows the recess in the base. This bullet does not expand to seal the bore as a lead core bullet will. The cavity in the base is there to achieve some measure of gas sealing and reduce barrel wear. For comparison, the base of a conventional lead core bullet is shown at right.

Cases:	Norma.
Primers:	Federal 215.
Bullet:	Steel tipped version of the turned brass bullet, weighing 165 gr.
Test results at 30 yards:	On a Birch tree some 14"-15" in diameter, this load penetrated both the tree and a 1/16" steel plate behind it. The hole in the plate was elliptical and 1/8" longer than it was wide. This version of the bullet is a little too long to be well stabilized by this rifle. The penetration is rather remarkable in that military FMJ bullets of 165, 172 and 173 gr. driven by the same powder charge failed to even penetrate the tree.
	The remaining energy of this bullet when it hit the steel plate was ample to put a man out of action.

TEST 2b

Rifle:	Same as in Test 2a.
Load:	Same as in Test 2a except bullet was a 162 gr. flat tipped brass bullet.
Test results at 30 yards:	This bullet went through the same tree as in Test 2a, but the hole in the 1/16" steel plate behind it was perfectly round and showed higher velocity when it hit the steel plate than load 2a did.

TEST 3a

Rifle:	.30-06 Winchester Model 70 as in Test 1.
Cases:	Arsenal.
Primers:	Federal 210.
Powder:	4350 – 55 gr.
Bullet:	Steel tipped manganese bronze, 178 gr. weight and 77 Rb hardness.
Velocity:	Not checked.
Test results at 200 yards:	On 1/2" steel plate, this bullet made a crater 3/16" deep x 5/8" in diameter. The bulge on the back of the plate was 1/16" high and covered an area as large as the crater. The pressure developed by this load was similar to the 190 gr. Barnes open base FMJ bullets.

TEST 3b

Rifle:	Same as in Test 3a.
Powder:	MR 3100 – 72 gr.

Cases:	Norma.
Primers:	Federal 215.
Bullet:	Steel tipped version of the turned brass bullet. 165 gr. weight.
Test results at 200 yards:	On car door A, this bullet penetrated the outer door panel, then both sides of the reinforcing panel with almost no evidence of expansion, blasted a special panel carrying a counterweight spring out of the way and finally hit the 1/2" backing plate at an angle. The crater was 1/16" deep and 1/2"x3/4" in size.
Comment:	This bullet was 1-7/16" long and was not completely stabilized. However, it showed very little energy loss in penetrating the car door and evidently arrived on the backing plate with most of its original weight.

15

Summary And Overview

The experiments presented in this book indicate that the design of military armor piercing bullets, as made by U.S. arsenals, has a strong basis for claiming to be well grounded on actual requirements. Arsenal-produced A.P. bullets have little tendency to skid upon striking a slanted surface. If correctly made, the bullets have a good ballistic coefficient and range well. When steel plate is struck, their design permits a great deal of penetration for the cartridge energy involved. If precisely made, the arsenal-produced A.P. is adequately accurate and it will inflict severe damage on an automobile engine. It has some disadvantages, too. The jacket ruptures rather easily and when it does, the core may tumble if any yaw is present in the bullet. Consequently, it will penetrate very little on multiple barriers. It does not penetrate trees well at all. It is quite long and is only marginally stabilized by a 10" twist barrel from a .30-06 or even a .308 Norma Magnum. The bullet requires a lot of expensive and very heavy machinery to manufacture.

The various alternate designs evaluated and written up in this text also have strong and weak points. From the standpoint of cost effectiveness, the heads-up winner is the altered soft point or hollow point hunting bullet with the front portion cut off and a shallow cavity drilled in the front and a steel tip, pressed into place. This bullet is similar to the

Remington Bronze Point in external appearance. As detailed earlier in the text, the jacket is beveled to match a corresponding bevel in the steel tip. This is done to greatly retard expansion. These tests used a Hornady bullet which has a specially made jacket to insure good expansion at long range and yet not blow up at short range. For hunting purposes, this is a great design, but it is not the best for our needs. The heaviest-made jacketed bullets readily available are made by Barnes Bullets of American Fork, Utah. In some calibers, a jacket .049" thick is available and this does produce a strong, stiff jacket. The contour of the 190 gr. Barnes special bullets allowed them to penetrate trees well but their open base made them fragile. The normal Barnes bullet has a closed base that is very strong and has a slightly greater thickness of jacket at the point than over the body. For our purposes, this is ideal. For the chap with only a small lathe, the only choice that makes any sense is the modified sporting bullet. The fine accuracy built into these bullets deteriorates very little if the tip alteration is done precisely and correctly. From a powerful rifle, such a bullet will penetrate most car fender material. Unless a boss on the engine block is struck, such a bullet will do severe disabling damage to a car engine with a single hit from a rifle of the .30-06's power level if the range does not extend beyond 300 yards. Steel penetration capability is above the standard soft point or FMJ military bullet of equivalent size. I recommend a long, flat base design for alteration in preference to a boattail design bullet. The longer bearing surface of the flat based bullet means better wood penetrating ability.

The various experiments detailed in this book show the turned manganese bronze bullet has good wood penetrating ability and this design is also good even if merely made out of brass in the 60-65 Rockwell B range. Accuracy of the steel core special bullet was compromised by my driving arrangement, but if made with the 3 wide bands it can be very accurate. The best way to make them in quantity would be to use commercial jackets and cut them to length, then press the jacket over the rear shank of the bulle. It may be necessary to drill a very small hole dead center in the base of

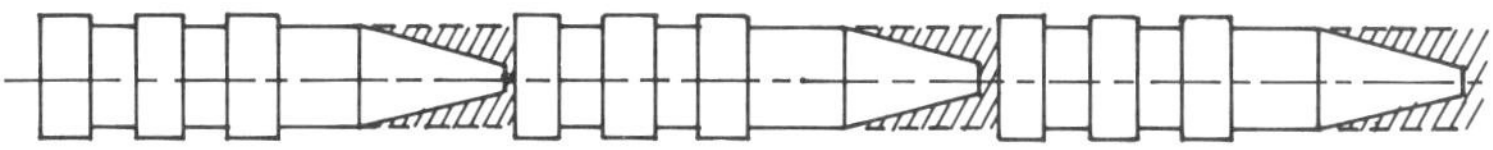

FIGURE 3

The rod for lathe turned bullets as it would come from the infeed centerless grinder with the bands and grooves ground to the correct diameter is shown above. The cavity in the base is cut on the first lathe operation and the shaded areas are removed in the second operation to form the point.

the bullet jacket to prevent air being trapped in the jacket during the assembly procedure. The core of the bullet should be at least 45 Rockwell C and 50 would be better. This means a grinder is needed to shape the core, in most cases, unless a very rigid lathe with carbide tooling is used. As a core material, steel has some severe drawbacks. For one thing, it is lighter than lead and this lower density is a disadvantage for steel penetration. Higher density material, like tungsten carbide, requires diamond grinding wheels for shaping, although some varieties use silicon carbide grinding wheels. If one had access to a shop making drills from tungsten carbide and could purchase defective drills for scrap prices and grind the shanks into bullet cores, a phenomenally efficient A.P. bullet would result. An alternate idea is Stellite in a machineable grade. I bought a 1/4" rod, 12" long and paid nearly a dollar an inch for it. There is little doubt that it would make an excellent A.P. bullet. However, the expense is enough to give pause to any but a well-heeled survivalist or a well-financed police department. There is a real opportunity for a grinding shop to furnish ground and grooved hard steel rods 6" or so long to the lathe owning hobbyist or police department. For quantity production, the Stellite core could be ground for the same type of alterations. The extra density of the Stellite would make a heavier bullet.

These experiments also indicate a dead sharp point is undesirable on an A.P. bullet that lacks anything ahead of the steel nose. A slightly rounded tip or even a flat tip some 1/16" in diameter is better as it reduces the tendency to skid on a slanted surface. A soft steel core is easy to make but is of very little use on thick steel plates and is only fair in certainty for damaging a car engine.

The very thick jacketed round nosed FMJ bullet used on thick skinned African game had good points. Such bullets are very strong and plow through fender type material with ease. They are also very destructive on engine blocks. Additionally, they have phenomenal penetration on trees. Their only real drawback is a low ballistic coefficient and consequently reduced energy at long range, combined with inferior trajectory and higher wind drift as compared to a Spitzer bullet. If these bullets had a semi-Spitzer shape, they would come close to being an ideal bullet for a lot of sniper rifle situations.

These bullets pretty much showed up the good and bad points of high velocity on diverse targets with a variety of bullets. The structural weakness of commercial bullets for sniper rifle usage was documented on a variety of targets. How to design a two-piece bullet for good and bad accuracy due to driving band differences was discovered and documented. This book does not pretend to be the final word on the subject but enough information has been developed to prevent a police officer, survivalist or security minded individual from making erroneous assumptions and wasting time and money and in some cases even risking their lives.

16

Summary Of Test Results

The following tests were conducted on several types of barriers and provided some unexpected results. The barriers included: Steel plates, car doors, trees, concrete blocks and an automobile engine block.

The steel plate barrier test will be summarized first. The variables include the following:

1. Bullet core hardness.
2. Bullet core density.
3. Bullet impact velocity.
4. Bullet weight.
5. Bullet structure and internal strength.

Core hardness variation was most dramatically shown when armor piercing projectiles were compared to lead core bullets of similar weight, identical diameter and closely similar velocities. In both .30-06 and .308 Norma Magnum, the 165 gr. A.P. had twice the penetration capability of lead core bullets of 172 and 180 gr. weight on steel plate.

The third variable, impact velocity, was extremely important on lead core bullets. For example, the 180 gr. Hornady Spire Point with a muzzle velocity of 2450 fps from a .308 Winchester failed to penetrate the 3/8" plate at 200 yards. The same bullet, traveling at some 200 fps faster from a

High sectional density means a long bullet. From left to right: Hornady 175 gr., .284" diameter Spire Point with a sectional density value of 310; Sierra 200 gr., .308" diameter Spitzer Boattail with a sectional density value of 301; Hornady 180 gr., .308" diameter Spire Point with a sectional density value of 270.

.30-06, did penetrate the 3/8" steel plate. The real proof of the significance of velocity is shown when the .308 Norma Magnum discharged the same bullet at 3100 fps. They sailed right through the 1/2" steel plate at 200 yards. Equally dramatic was the fact that the 60 gr. Hornady bullet, traveling at about 3500 fps, passed right through the 3/8" plate at 200 yards while the 180 gr. Hornady in a .308 Winchester failed.

Bullet weight proved to be a factor to be reckoned with. The 270 gr. Hornady Spire Point, fired from the .375 H&H, had a muzzle velocity only slightly higher than the .30-06 173 and 180 gr. bullets. But the .375 H&H would penetrate the 1/2" plate at 200 yards and the lighter bullets from the .30-06 would not. The 250 gr. bullet, fired from the .33 Wildcat on the .30-06 case, had almost identical impact velocity as the 180 gr. .30-06 bullets. Yet, the larger bullet came closer to penetrating the 1/2" plate at 200 yards.

Bullet structure did not seem to matter much as long as lead core bullets were used. There was some difference between Hornady and Sierra bullet craters on steel plate, however. The Sierra bullet craters tended to be larger in diameter and shallower than those made by Hornady bullets. The military 173 gr. Spitzer Metal Case Boattail bullets performed very similarly to the 180 gr. Hornady Spire Point on steel plate. The modified 180 gr. Hornady Spire Point bullets with a tool steel tip would penetrate the 1/2" plate at 200 yards when fired from a .30-06.

The car door tests showed that bullet structure, which is primarily a function of jacket material, jacket thickness and point design, was extremely important. The 6.5mm 139 gr. FMJ Norma bullet, with its steel jacket and Spitzer full metal case design, did relatively well on the car door tests. The 8x57 Mauser 154 gr. FMJ bullet with a very heavy jacket also did fairly well on the car door. The .33 caliber Barnes bullets did reasonably well for a soft point bullet so long as velocity was compatible with the jacket thickness.

Bullet impact velocity had different effects on FMJ and soft point type bullets. The FMJ bullets seemed to penetrate better at higher velocities. The soft point bullets came apart faster when muzzle velocity increased from 2050 fps to 3100 fps.

Lead core bullets of conventional size and construction are just too fragile to get through any part of a car door with 100% reliability. If only the outer and inner panels are hit, there is little problem in penetration. But if the reinforcing panel gets in the way, reliable penetration is hard to achieve.

Military A.P. was a big disappointment. When the core was concentric with the jacket, results were fairly good. Most of those that I had tested were terrible in the accuracy department and tipped after hitting the outer panel. After tipping, their path was very unpredictable.

The steel tipped Hornady Spire Point and the lathe turned homogenous bullets were fair on the car door test but not dramatically efficient.

In pointed .30 caliber soft point bullets, the Sierra proved to be more fragile than the Hornady. The very rapid expansion of the Nosler bullets prevented them from being impressive on this test.

The tests on trees and logs were eye openers. Soft point hunting bullets will penetrate only a few inches of wood. Military FMJ bullets apparently tumble after impact, when of the conventional Spitzer design. The old round nosed FMJ bullets used in the .30-40 Krag and the .30-03 Springfield would penetrate trees beautifully. The military A.P. 165 gr. bullets were poor on wood penetration in these tests.

The only lead core bullet, of any that I tested, that had any decent penetration in wood was the .30 caliber 190 gr. Barnes with an open base, a long cylindrical bearing area and a point of the semi-Spitzer shape but which was fairly short in length. They were made from .032" thick cups and had a closed point and an open base. The homogenous lathe turned bullets were very good on wood penetration, but adding a steel tip appeared to add too much to the length of the bullet and decreased stability. The long cylindrical area kept if from tumbling. The round nose FMJ bullets made by Hornady and Barnes will penetrate an incredible amount of wood or trees.

The .223 FMJ bullet used in Vietnam was a very poor design for penetrating brush or trees. It had marginal stability, low mass, a long Spitzer point, a very short cylindrical section and rather fragile construction.

The survival conscious individual, who has a retreat surrounded by trees a foot or more in size, had better get some round nosed FMJ bullets or lathe turned homogenous bullets for his battle rifle reloads. I would personally favor manganese bronze over brass for the bullets, but 77 Rb is as hard as I would care to use. These slugs would take very good care of some hostile looter hiding behind a tree and spraying lead with a sub-machine gun.

Concrete blocks require a strongly made bullet, considerable bullet mass and a high level of kinetic energy to really take them apart. The 200 gr. .30 caliber Sierra Soft Point Boattail bullet, from the .30-06, would shatter on the first layer of the block and spray the second layer with ground-up bullet particles and fine bits of concrete. Oddly enough, the 55 gr. .223 bullet did just about the same thing. The 173 gr. military FMJ Spitzer Boattail appeared to penetrate better from the 12" twist of the .308 Winchester than from the 10" twist of the .30-06.

The 190 gr. FMJ Barnes .30 caliber bullet performed rather poorly on concrete blocks. This bullet has an open base and this appears to detract from the structural rigidity needed on concrete blocks.

The .308 Norma Magnum with the 180 gr. Hornady Spire Point at 3100 fps was quite destructive on concrete blocks. The block was shattered into several pieces when this load was used. Rifles, with muzzle energies in the 2800 to 3000 fps range, would damage a block badly and/or break it into pieces. They did not shatter the blocks as the more powerful rifles had done. The 7mm/06 and .30-06 using 175 and 180 gr. Hornady Spire Point bullets respectively, were in this class. What edge there was belonged to the .30-06, in spite of its slightly lower velocity and energy.

The .33 caliber rifles with 250 gr. Barnes bullets were extremely effective on concrete blocks. Both exceeded the .308 Norma Magnum in destructiveness on concrete blocks and the larger rifle did so by an impressive margin. I wish the .375 H&H could have been used in this test but scope problems prevented it.

Anyone who expects concrete blocks to present much of a barrier to someone bent on homicide with a powerful rifle is foolish in the extreme. Even the .223 will go through, if 2 or 3 shots hit the same place.

The car engine block test presented its own share of surprises. The core hardness and structural strength of a bullet are very important if an engine block is going to be damaged enough to immobilize the vehicle. The ribs, bosses and similar reinforcing points of an engine block cause an enormous variation in its ability to withstand a bullet's impact. The 270 gr. Hornady Spire Point, fired from the .375 H&H, hit a very heavily reinforced part of the engine and did very little damage. On the other hand, both the .222 Remington and .223 Remington penetrated the water jacket at a thin point.

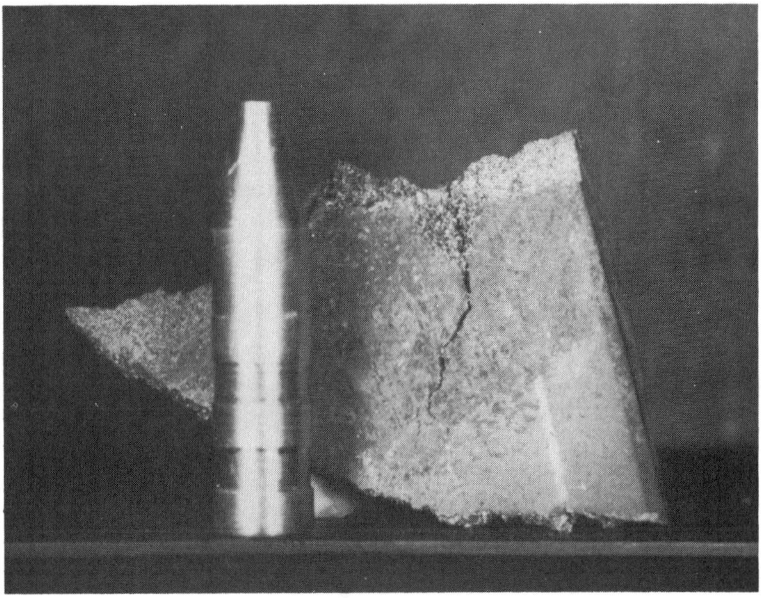

The chunk of metal knocked out of an engine block and a lathe turned bullet identical to the one that knocked it out.

The 300 gr. Hornady FMJ, made for the .375 H&H with an extremely tough jacket, proved very capable as an engine demolisher. The modified 180 gr. Hornady Spire Point, when given a steel tip and fired from the .308 Norma Magnum, also was an engine wrecker. The 165 gr. A.P. bullet, when fired from a .308 Norma Magnum, would also be classified as effective on destroying engines. The hard core of this bullet gave it a great advantage over any lead core bullet. The incredibly poor accuracy of the majority of those I tested rendered them quite unreliable for the task at hand, however. If an accurate lot of armor piercing could be found, it would be one of the more desirable loads to use against a car engine.

The brass bullets made on a lathe were relatively good at producing engine damage. They are softer than an A.P. core but a great deal harder than any lead core bullet.

If I were to buy a commercial rifle for the express purpose of ending the operation of a car's engine, I would get a .375

H&H or .378 Weatherby and load 300 gr. Hornady FMJ bullets at as high a speed as I could get. Such a rifle would be very useful against powered boats operated by pirate gangs. The bullet would penetrate the hull and still be in shape to tear up the motor.

If these FMJ bullets had a semi-Spitzer shape and equal toughness, they would retain velocity much better and be a close approximation to the ideal combat bullet for smashing car or truck engines out to a relatively long range.

Where very long ranges are involved, a heavy weight, thick jacketed, steel tipped bullet would be about the best choice. The 200 gr. Barnes .30 caliber Spitzer and the Sierra .375 caliber, 300 gr. Spitzer would be good bullets for such alterations.